RANDOM ALLOCATIONS

SCRIPTA SERIES IN MATHEMATICS

Tikhonov and Arsenin · Solutions of Ill-Posed Problems, 1977

Rozanov · Innovation Processes, 1977

Pogorelov · The Minkowski Multidimensional Problem, 1978

Kolchin, Sevast'yanov, and Chistyakov · Random Allocations, 1978

Boltyanskiy · Hilbert's Third Problem, 1978

RANDOM ALLOCATIONS

VALENTIN F. KOLCHIN
BORIS A. SEVAST'YANOV
VLADIMIR P. CHISTYAKOV
Steklov Institute of Mathematical Sciences, Moscow

Translation Editor

A. V. BALAKRISHNAN
University of California, Los Angeles

1978
V. H. WINSTON & SONS
Washington, D.C.

A HALSTED PRESS BOOK

JOHN WILEY & SONS
New York Toronto London Sydney

V. H. Winston & Sons, a Division of Scripta Technica, Inc., Publishers
1511 K St. N.W., Washington, D.C. 20005

Distributed solely by Halsted Press, a Division of John Wiley & Sons, Inc.

Library of Congress Cataloging in Publication Data
Kolchin, Valentin Fedorovich.
Random allocations.

(Scripta series in mathematics)
Translation of Sluchaĭnye razmeshcheniia͡.
Bibliography: p.
Includes index.
1. Combinatorial probabilities. 2. Stochastic processes. 3. Distribution (Probability theory)
I. Sevast'ia͡nov, Boris Aleksandrovich, joint author. II. Christia͡kov, Vladimir Pavlovich, joint author. III. Title. IV. Series.
QA273.45.K6413 519.2 77-27814
ISBN 0-470-99394-4

Composition by **Isabelle Sneeringer**, Scripta Technica, Inc.

CONTENTS

FOREWORD

Combinatorial problems and methods play an important role in the theory of probability. Chief among these, in the literature devoted thereto, are: combinatorial problems in the theory of stochastic processeses, problems associated with random mappings and random graphs, and problems of allocations of particles to cells. During the last decade, considerable effort has been concentrated on allocation problems. Among these, the classical problem of shots is perhaps the best known. Let n shots be thrown independently into N cells. What is the distribution of the random variable $\mu_0(n, N)$ denoting the number of empty cells?

This problem can be given many different interpretations.

Consider for example a random mapping of a set of n points into itself, in which each of the n points is mapped into one of the points of the set randomly and independently. This mapping can be associated with a random graph with n vertices each of which is connected by an arc to one of the vertices in a random way. The variable $\mu_0(n, n)$ is then the number of vertices of zero multiplicity in this case (i.e., having no preimage).

Other formulations of this and related problems are also known.

For example, the number of days in a year which are not birthdays of n randomly chosen people ($\mu_0(n, 365)$); the number of stamps or postcards a collector must acquire to make a complete set of N pieces, assuming independent, random acquisition every time.

The recent interest in such combinatorial problems is due on the one hand to the availability of high-speed digital computers and on the other hand to the development of asymptotic techniques. Random allocation problems occur in mathematical statistics, theory of automata, statistical physics, computer science, communication networks, astronomy, biology, etc. They also provide a fruitful area of application of many asymptotic methods of probability theory: the method of moments, the method of saddle-points, "companion-scheme" methods (where a complex scheme is replaced with a similar but simpler scheme), the method of reduction to sums of independent and/or dependent random variables, the method of reduction to conditional distributions of independent random variables, and others.

Simplicity of problem formulation, combined with the richness and variety of the techniques of solution, make allocation problems good vehicles for demonstrating asymptotic methods of probability theory in the class-room. The results presented here have been developed mainly during the last decade.

The book consists of 8 chapters. Chapters 1 and 2 deal with equiprobable allocations. A whole spectrum of limit theorems is obtained for the number of empty boxes $\mu_0(n, N)$ and the number of boxes $\mu_r(n, N)$ with fixed occupancy r. Chapter 3 contains generalizations of the results of the first two chapters to multinominal allocation. In Chapter 4, the convergence to Poisson and Gaussian processes is proved, when the boxes are filled by n balls sequentially. In Chapter 5, the limit theorems in Chapters 2 and 3 are used to determine the optimal statistic for the empty cell test and its generalization. The remaining chapters deal with some generalized schemes of allocation; in particular, Chapter 6 treats allocations when the number of particles is random, and Chapter 7 deals with allocation of particles by clumps ('complexes'). In Chapter 8, a generalized allocation scheme is used to study cycles of random permutations.

The book can be understood by readers familiar with standard courses in mathematics and probability theory. References are given in the last section of each chapter. A list of references on random allocations up to 1973 can be found in Kolchin and Chistyakov (34) published by VINITI in the series ITOGI NAUKI (Russian).

The scheme used for numbering sections, formulas, theorems, lemmas, corollaries and notes is as follows. Each chapter has its own numbering of sections, and each section its own numbering of formulas, theorems, etc. For citations within the same section, only the entry number is used. In referring within the same chapter to the formulas, theorems, etc., in a different section, the number of the section is affixed in front. In referring to a section, theorem, formula, etc., of a different chapter, the number of the chapter is affixed in front. For example, Section 2.3 means Section 3, Chapter 2; formula (3.5) means formula 5 of Section 3 of the same chapter where the reference is made. Formula (8.3.5) means formula 5 in Section 3 of Chapter 8. Similarly, we refer to Theorem 2 of Section 3 of Chapter 4 as Theorem 2 in the same section, as Theorem 3.2 in the other sections of Chapter 4, and as Theorem 4.3.2 in other chapters.

The writing was split up among the three authors in the following way: Kolchin, Chapters 2 and 8; Sevast'yanov, Chapters 1, 2, 3, 4, 6, and 7; Chistyakov and Kolchin, Chapters 1, 2, 3, and 5.

We take this opportunity to thank the editor Andrei Mikhailovich Zubkov whose comments and suggestions helped to improve the presentation and correct some errors.

V. F. Kolchin,
B. A. Sevast'yanov,
V. P. Chistyakov

Chapter I

THE CLASSICAL SHOT PROBLEM

§1. The Number of Empty Cells

To verify the statistical hypothesis H_0 that a sample of independent observations $x_1, x_2, \ldots, x_n$, is taken from a continuous distribution $F(x)$, the so-called "empty cell" test is sometimes used. With it, points $z_0 = -\infty < z_1 < \ldots < z_{N-1} < z_N = \infty$ are chosen so that $F(z_k) - F(z_{k-1}) = 1/N$, for $k=1, 2, \ldots, N$. The test is based on the statistic μ_0, which is the number of intervals $(z_{k-1}, z_k]$, containing no observation x_i. The test is defined as follows: if $\mu_0 \leqslant C$, the hypothesis H_0 is adopted; if $\mu_0 > C$, the hypothesis H_0 is rejected. The constant C is chosen so that the error of the first kind, i.e., the probability of rejecting the hypothesis H_0, when it is true, is equal to a given fixed value.

The empty cell test is attractive because of its simplicity. It is computationally simpler than others (for example, Kolmogorov's criterion or the χ^2-test) a fact which is crucial for its many applications. The problem of computing the empty cell test is related to the following classical shot problem.

Let there be N boxes. Let n shot be thrown into these boxes randomly and independently. We assume that the probability for any particular shot to fall into the jth box is $1/N$ for $j=1, 2, \ldots, N$. Let us denote by $\mu_0=\mu_0(n, N)$ the number of empty boxes. It can readily be seen that the distribution of this variable is the same as that of the statistic on which the empty cell test is based.

The problem of the number of empty boxes in the scheme described above can be a model for many real phenomena in various branches of science, engineering, and other fields of applications of probability theory and mathematical statistics. The empty cell test can prove useful in verifying the hypothesis of uniform distribution of objects in a bounded space (for example, stars in a part of outer space, individuals of a biological population in a given area, etc.). In designing a complex system one sometimes needs to make simplifying assumptions leading to a plan of allocation of shots to boxes or, equivalently, allocation of particles to cells.

The law of distribution of the number of empty boxes $\mu_0(n, N)$ is given by the formulas

$$\mathbf{P}\{\mu_0(n, N) = k\} = C_N^k\left(1 - \frac{k}{N}\right)^n \mathbf{P}\{\mu_0(n, N-k) = 0\}, \quad (1)$$

$$\mathbf{P}\{\mu_0(n, N) = 0\} = \sum_{l=0}^{N} C_N^l (-1)^l \left(1 - \frac{l}{N}\right)^n. \quad (2)$$

We shall prove (1) and (2). Let A_i be the event that the ith cell is left empty, and let $\bar{A}_i$ be the event complementary to A_i. Then

$$\mathbf{P}\{\mu_0(n, N) = k\} = \sum_{1 \leq i_1 < \ldots < i_k \leq N} \mathbf{P}\left\{A_{i_1} \ldots A_{i_k} \bar{A}_{j_1} \ldots \bar{A}_{j_{N-k}}\right\}, \quad (3)$$

where $\{j_1, \ldots, j_{N-k}\} = \{1, 2, \ldots, N\} \setminus \{i_1, \ldots, i_k\}$. All the terms in (3) are the same and the number of them is C_N^k. By the formula for joint probabilities:

$$\mathbf{P}\left\{A_{i_1}\dots A_{i_k}\overline{A}_{j_1}\dots\overline{A}_{j_{N-k}}\right\}=$$

$$=\mathbf{P}\{A_{i_1}\dots A_{i_k}\}\,\mathbf{P}\left\{\overline{A}_{j_1}\dots\overline{A}_{j_{N-k}}\,\middle|\,A_{i_1}\dots A_{i_k}\right\},$$

where

$$\mathbf{P}\{A_{i_1}\dots A_{i_k}\}=\left(1-\frac{k}{N}\right)^n \tag{4}$$

and $\mathbf{P}\left\{\overline{A}_{j_1}\dots\overline{A}_{j_{N-k}}\,\middle|\,A_{i_1}\dots A_{i_k}\right\}=\mathbf{P}\{\mu_0(n,N-k)=0\}$, from which (1) follows. To prove (2), we need to use the formula for the probability of the sum of events

$$\mathbf{P}\{\mu_0(n,N)>0\}=\mathbf{P}\left\{\bigcup_{i=1}^{N}A_i\right\}=$$

$$=\sum_i\mathbf{P}(A_i)-\sum_{i<j}\mathbf{P}(A_iA_j)+\sum_{i<j<k}\mathbf{P}(A_iA_jA_k)-\dots,$$

from which formula (2) follows by virtue of (4). To carry out the computations, it is sometimes more convenient to use not (1) and (2) but the recurrence formula

$$\mathbf{P}\{\mu_0(n+1,N)=k\}=$$

$$=\left(1-\frac{k}{N}\right)\mathbf{P}\{\mu_0(n,N)=k\}+\frac{k+1}{N}\mathbf{P}\{\mu_0(n,N)=k+1\}, \tag{5}$$

which can be proved with the help of the joint probability formula.

Formulas (1), (2), and (5) are inconvenient for analyzing the distribution of $\mu_0(n, N)$ for large n and N. Hence, other methods of investigating the characteristics of $\mu_0(n, N)$ are of greater interest. The moments of $\mu_0(n, N)$ can be obtained from the equation

$$\mu_0(n, N)=\theta_1+\theta_2+\dots+\theta_N, \tag{6}$$

where $\theta_i=1$ if the ith box is left empty and zero otherwise. We have

$$\mathbf{M}\mu_0(n, N) = \sum_{i=1}^{N} \mathbf{M}\theta_i ,$$

and since $\mathbf{M}\theta_i = \mathbf{P}\{\theta_i = 1\} = \left(1 - \frac{1}{N}\right)^n,$ it follows that

$$\mathbf{M}\mu_0(n, N) = N\left(1 - \frac{1}{N}\right)^n. \tag{7}$$

Let us introduce the notation for factorial powers:

$$x^{[k]} = x(x-1)\ldots(x-k+1), \qquad x^{[0]} = 1. \tag{8}$$

Since $\mu_0^{[2]} = \mu_0^2 - \mu_0$ and $\theta_i^2 = \theta_i$, we have $\mu_0^{[2]} = \sum_{i \neq j} \theta_i \theta_j$, from which it follows that

$$\mathbf{M}\mu_0^{[2]} = \sum_{i \neq j} \mathbf{M}\theta_i \theta_j = \sum_{i \neq j} \mathbf{P}\{\theta_i = \theta_j = 1\}$$

and
$$\mathbf{M}\mu_0^{[2]} = N(N-1)\left(1 - \frac{2}{N}\right)^n. \tag{9}$$

In the general case, for any positive integer k,

$$\mathbf{M}\mu_0^{[k]} = \sum_{i_1 \ldots i_k}^{*} \mathbf{P}\{\theta_{i_1} = \ldots = \theta_{i_k} = 1\}, \tag{10}$$

where the summation $\sum^*$ is over all $N^{[k]}$ ordered sets $(i_1, \ldots, i_k)$ with pairwise unequal indices. We have from (10)

$$\mathbf{M}\mu_0^{[k]} = N^{[k]}\left(1 - \frac{k}{N}\right)^n. \tag{11}$$

From (7) and (9), we can calculate the variance:

$$\mathbf{D}\mu_0 = N(N-1)\left(1-\frac{2}{N}\right)^n + N\left(1-\frac{1}{N}\right)^n - N^2\left(1-\frac{1}{N}\right)^{2n}. \tag{12}$$

Theorem 1. *For all n, N, we have the inequality*

$$\mathbf{M}\mu_0(n,\ N) \leqslant Ne^{-\alpha}, \tag{13}$$

where $\alpha = n/N$. Also, if $n,\ N \to \infty$ in such a way that $\alpha = o(N)$, then

$$\mathbf{M}\mu_0 = Ne^{-\alpha} - \frac{\alpha}{2}e^{-\alpha} + O\left(\frac{\alpha(1+\alpha)e^{-\alpha}}{N}\right) \tag{14}$$

and

$$\mathbf{D}\mu_0 = Ne^{-\alpha}(1-(1+\alpha)e^{-\alpha}) + O\left(\alpha(1+\alpha)e^{-\alpha}\left(e^{-\alpha}+\frac{1}{N}\right)\right). \tag{15}$$

PROOF. Inequality (13) follows from (7) and the inequality $1-\frac{1}{N} < e^{-1/N}$. Assertion (14) follows from (7) if we use the Taylor expansion: $\ln(1-x) = -x - \frac{x^2}{2} + O(x^3)$:

$$\begin{aligned}\mathbf{M}\mu_0 = N\left(1-\frac{1}{N}\right)^n &= N\exp\left(-\frac{n}{N} - \frac{n}{2N^2} + O\left(\frac{n}{N^3}\right)\right) = \\ &= Ne^{-\alpha}\left(1-\frac{\alpha}{2N} + O\left(\frac{\alpha+\alpha^2}{N^2}\right)\right) = \\ &= Ne^{-\alpha} - \frac{\alpha}{2}e^{-\alpha} + O\left(\frac{\alpha(1+\alpha)}{N}e^{-\alpha}\right).\end{aligned} \tag{16}$$

Similarly, we deduce (15) from (9), (12), and (16). By substituting the asymptotic expansions

$$\mathbf{M}\mu_0^{[2]} = (N^2 - N)\exp\left\{-n\left(\frac{2}{N} + \frac{2}{N^2} + O\left(\frac{1}{N^3}\right)\right)\right\} =$$
$$= (N^2 - N)\,e^{-2\alpha}\left(1 - \frac{2\alpha}{N} + O\left(\frac{\alpha + \alpha^2}{N^2}\right)\right) =$$
$$= N^2 e^{-2\alpha} - Ne^{-2\alpha}(1 + 2\alpha) + O\left(\alpha(1 + \alpha)\,e^{-2\alpha}\right),$$
$$(\mathbf{M}\mu_0)^2 = N^2 e^{-2\alpha} - N\alpha e^{-\alpha} + O\left(\alpha(1 + \alpha)\,e^{-2\alpha}\right)$$

into (12), we arrive at (15).

Using the asymptotic formulas (14) and (15), we can distinguish five different domains $n,\ N \longrightarrow \infty$, for which the asymptotic distribution $\mu_0(n,\ N)$ are different. The domain n, $N \longrightarrow \infty$ for which $0 < c_1 \leqslant \alpha \leqslant c_2 < \infty$, where c_1, c_2 are constants is said to be the *central domain.* In the central domain, $\mathbf{M}\mu_0$ and $\mathbf{D}\mu_0$ increase asymptotically in proportion to N and

$$\lim_{\frac{n}{N}\to\alpha} \frac{\mathbf{D}\mu_0}{\mathbf{M}\mu_0} = 1 - (1 + \alpha)\,e^{-\alpha} < 1.$$

If

$$\alpha \longrightarrow \infty \text{ and } \mathbf{M}\mu_0 \longrightarrow \lambda < \infty, \tag{17}$$

we shall say that n, N, or α varies in the *right-hand domain.* In this case, we have $\mathbf{D}\mu_0 \longrightarrow \lambda$ by virtue of (15). It follows from (17) that

$$\alpha = \ln N - \ln \lambda + o(1). \tag{18}$$

By the *left-hand domain* we mean those $n,\ N \longrightarrow \infty$ for which

$$\frac{n^2}{2N} \to \lambda < \infty. \tag{19}$$

In the left-hand domain, $\alpha \longrightarrow 0$, and it follows from (15) and (14) that

$$\mathbf{D}\mu_0 \sim N\frac{\alpha^2}{2} \to \lambda, \tag{20}$$

$$\mathbf{M}\mu_0 = N - n + \frac{N\alpha^2}{2} + o(1) = N - n + \lambda + o(1). \tag{21}$$

The number of particles in the left-hand domain is small compared to the number of cells. Hence, almost every particle gets in a cell by itself. If each of the nonempty cells contained one particle, the number of empty cells would be $N—n$. By virtue of (20) and (21), the excess of the number of empty cells over the level $N - n$, namely, $\mu_0—(N—n)$ in the left-hand domain has the same limiting variance and mathematical expectation.

Let us define the *left-hand intermediate domain* by the relations

$$\alpha \to 0, \qquad \frac{N\alpha^2}{2} \to \infty. \tag{22}$$

Under these conditions, $\mathbf{M}(\mu_0 - N + n) \longrightarrow \infty$, $\mathbf{D}\mu_0 \longrightarrow \infty$, and $\frac{\mathbf{D}\mu_0}{\mathbf{M}\mu_0} \to 0$.

Let us define the *right-hand intermediate domain* by the relations

$$\alpha \longrightarrow \infty, \qquad \mathbf{M}\mu_0 \longrightarrow \infty. \tag{23}$$

In such a domain, $\alpha - \ln N \longrightarrow -\infty$, $\mathbf{D}\mu_0 \longrightarrow \infty$, and $\frac{\mathbf{D}\mu_0}{\mathbf{M}\mu_0} \to 1$.

We shall see below that corresponding to these five kinds of domain we have different limit theorems for μ_0. We shall use the simple method of moments to prove

Theorem 2. *In the right-hand domain, we have*

$$\lim_{N\to\infty} \mathbf{P}\{\mu_0(n, N) = k\} = \frac{\lambda^k}{k!}e^{-\lambda}, \quad k = 0, 1, 2, \ldots, \tag{24}$$

where $\lambda = \lim \mathbf{M}\mu_0$.

PROOF. Substituting (17) into (11), we have

$$\mathbf{M}\mu_0^{[k]} = N^{[k]}\left(1 - \frac{k}{N}\right)^n \sim N^k e^{-kn/N} = N^k e^{-\alpha k} \to \lambda^k.$$

The limiting values of the factorial moments $\mathbf{M}\mu_0^{[k]}$ coincide with those of a Poisson factorial distribution. The Poisson distribution is uniquely determined by its moments and hence we have (24).

The theorem is proved.

The method of moments is convenient in considering sequences of noncentral and nonstandardized random variables. However, in investigating the convergence of the distributions of a sequence of integer-valued random variables to a normal distribution, we have to centralize and standardize first. In this case, the method of moments is less convenient; instead, one frequently uses characteristic functions, which can be readily obtained from the generating functions

$$F_{n,N}(x) = \mathbf{M}x^{\mu_0(n,N)}, \qquad |x| \leqslant 1. \tag{25}$$

Theorem 3. *Let us define*

$$\Phi^{(N)}(x, z) = \sum_{n=0}^{\infty} \frac{z^n N^n}{n!} F_{n,N}(x). \tag{26}$$

Then

$$\Phi^{(N)}(x, z) = (e^z + x - 1)^N. \tag{27}$$

PROOF. Let us divide the number of boxes into two groups consisting of N_1 and $N_2 = N - N_1$ boxes, respectively.

We can calculate the probability of the event $\{\mu_0(n, N) = k\}$ by considering the joint probability, fixing the numbers of shots n_1 and n_2 in the groups obtained:

$$\mathbf{P}\{\mu_0(n, N) = k\} = \sum_{n_1+n_2=n} \sum_{k_1+k_2=k} C_n^{n_1} \left(\frac{N_1}{N}\right)^{n_1} \left(\frac{N_2}{N}\right)^{n_2} \times$$
$$\times \mathbf{P}\{\mu_0(n_1, N_1) = k_1\} \mathbf{P}\{\mu_0(n_2, N_2) = k_2\}.$$

Let us multiply the two sides of this equation by $\frac{z^n N^n}{n!} x^k$ and $\frac{z^{n_1} z^{n_2} N^{n_1} N^{n_2}}{n!} x^{k_1} x^{k_2}$ respectively, and let us sum with respect to all n, k. We have

$$\Phi^{(N)}(x, z) = \sum_{n_1, n_2=0}^{\infty} \sum_{k_1, k_2=0}^{\infty} \frac{N_1^{n_1} N_2^{n_2}}{n_1! n_2!} z^{n_1} z^{n_2} x^{k_1} x^{k_2} \times$$
$$\times \mathbf{P}\{\mu_0(n_1, N_1) = k_1\} \mathbf{P}\{\mu_0(n_2, N_2) = k_2\}.$$

The summands in the right-hand side of this equation can be represented as the product of two factors:

$$\frac{N_l^{n_l}}{n_l!} z^{n_l} x^{k_l} \mathbf{P}\{\mu_0(n_l, N_l) = k_l\}, \qquad l = 1, 2,$$

and, therefore, the right-hand side can be written as the product of two series equal to $\Phi^{(N_1)}(x, z)$ and $\Phi^{(N_2)}(x, z)$ respectively. Thus,

$$\Phi^{(N)}(x, z) = \Phi^{(N_1)}(x, z) \Phi^{(N_2)}(x, z).$$

From this we obtain

$$\Phi^{(N)}(x, z) = [\Phi^{(1)}(x, z)]^N. \tag{28}$$

Since

$$F_{n,1}(x) = \mathbf{M} x^{\mu_0(n,1)} = \begin{cases} 1, & n > 0, \\ x, & n = 0, \end{cases}$$

we have

$$\Phi^{(1)}(x, z) = \sum_{n=0}^{\infty} \frac{z^n}{n!} F_{n,1}(x) = e^z + x - 1.$$

From this and (28) the assertion of the theorem follows.

Using the Cauchy formula, we obtain from (26) and (27) an expression for $F_{n,N}(x)$ as a contour integral:

$$F_{n,N}(x) = \frac{1}{2\pi i} \frac{n!}{N^n} \oint (e^z + x - 1)^N \frac{dz}{z^{n+1}}, \tag{29}$$

where the integral is over any closed contour encircling the point $z=0$. If we set $x=0$ in (29), we get

$$\mathbf{P}\{\mu_0(n, N) = 0\} = \frac{1}{2\pi i} \frac{n!}{N^n} \oint (e^z - 1)^N \frac{dz}{z^{n+1}}. \tag{30}$$

All factorial moments μ_0 can be computed by use of the formula

$$\mathbf{M}\mu_0^{[k]} = \left. \frac{d^k F_{n,N}(x)}{dx^k} \right|_{x=1}.$$

Let us set $\mathbf{M}\mu_0^{[k]} = m_k(n, N)$. Differentiating equation (26) k times at the point $x=1$ and taking (27) into account, we obtain

$$\sum_{n=0}^{\infty} \frac{z^n N^n}{n!} m_k(n, N) = N^{[k]} e^{z(N-k)} = \sum_{n=0}^{\infty} \frac{z^n}{n!} N^{[k]} (N-k)^n.$$

from which we obtain

$$m_k(n, N) = N^{[k]} \left(1 - \frac{k}{N}\right)^n.$$

This is the same result as (11), which was obtained by representing $\mu_0(n, N)$ in the form of the sum (6).

Let us note another useful fact. We shall introduce the random variables $\delta_1=1$, δ_l, $l=2, \ldots, N-1, N$, where δ_l is the number of shots thrown after $l-1$ boxes are filled and before l boxes are filled. The variables δ_l are independent and have geometric distributions:

$$\mathbf{P}\{\delta_l = m\} = \left(\frac{l-1}{N}\right)^{m-1}\left(1-\frac{l-1}{N}\right), \quad l = 2, \ldots, N. \tag{31}$$

The random variable

$$\nu_k=\delta_1+\delta_2+\ldots+\delta_k \tag{32}$$

is the smallest number representing the trial at which exactly k boxes become filled. It can be easily verified that

$$\mathbf{P}\{\nu_k\leqslant n\}=\mathbf{P}\{\mu_0(n, N)\leqslant N-k\}. \tag{33}$$

Equation (33) allows us to reduce the investigation of either random variable to that of the other.

§2. The Number of Trials before k Boxes Can be Filled

Let us consider the asymptotic behavior of the variable ν_k, equal to the smallest number of shots that will cause k boxes to be occupied, as $N \longrightarrow \infty$. The variable ν_k can be represented as the sum (1.32) of independent random variables δ_l having the distributions (1.31). The generating function of δ_l is

$$\mathbf{M}x^{\delta_l} = \left(1-\frac{l-1}{N}\right)\sum_{m=1}^{\infty}\left(\frac{l-1}{N}\right)^{m-1} x^m = \left(1-\frac{l-1}{N}\right)\frac{x}{1-\frac{l-1}{N}x}. \tag{1}$$

Differentiating (1) k times with respect to x and setting $x=1$, we get

$$\mathbf{M}\delta_l^{[k]} = k!\,\frac{N}{l-1}\left(\frac{l-1}{N-l+1}\right)^k, \quad l \geqslant 2. \tag{2}$$

In particular,

$$\begin{aligned} &\mathbf{M}\delta_l = \frac{N}{N-l+1}, \quad \mathbf{M}\delta_l^{[2]} = \frac{2N(l-1)}{(N-l+1)^2},\\ &\mathbf{D}\delta_l = \frac{N^2}{(N-l+1)^2} - \frac{N}{N-l+1}, \quad 1 \leqslant l \leqslant N. \end{aligned} \tag{3}$$

From this for $\nu_k = \delta_1 + \delta_2 + \ldots + \delta_k$ we get

$$\mathbf{M}\nu_k = N \sum_{l=N-k+1}^{N} \frac{1}{l}, \quad \mathbf{D}\nu_k = N^2 \sum_{l=N-k+1}^{N} \frac{1}{l^2} - \mathbf{M}\nu_k. \tag{4}$$

Let us find asymptotic formulas for $\mathbf{M}\nu_k$ and $\mathbf{D}\nu_k$ as k, $N \longrightarrow \infty$.

Theorem 1. *If* $k = N - c$, *where* $c \geqslant 0$ *is a fixed integer, then*

$$\mathbf{M}\nu_k = N \ln N + O(N), \quad \mathbf{D}\nu_k = d_c N^2 - N \ln N + O(N), \tag{5}$$

as $N \longrightarrow \infty$, *where*

$$d_0 = \frac{\pi^2}{6}, \quad d_c = \frac{\pi^2}{6} - \sum_{n=1}^{c} \frac{1}{n^2}, \quad c \geqslant 1.$$

Proof. Let us denote by $[x]$ the largest integer not exceeding x. Integrating the inequalities $x^{-1} \leqslant [x]^{-1} \leqslant (x-1)^{-1}$ from $N-k+2$ to $N+1$, we have

$$\ln(N+1) - \ln(N-k+2) \leqslant \sum_{l=N-k+2}^{N} \frac{1}{l} \leqslant \leqslant \ln N - \ln(N-k+1). \tag{6}$$

The first formula in (5) follows from (6) and (4). The second formula in (5) follows from (4) and the equations

$$\sum_{l=c+1}^{N} \frac{1}{l^2} = \sum_{l=1}^{\infty} \frac{1}{l^2} - \sum_{l=1}^{c} \frac{1}{l^2} - \sum_{l=N+1}^{\infty} \frac{1}{l^2},$$

$$\sum_{l=1}^{\infty} \frac{1}{l^2} = \frac{\pi^2}{6}, \quad \sum_{l=N+1}^{\infty} \frac{1}{l^2} = O\left(\frac{1}{N}\right), \quad \sum_{l=1}^{0} \frac{1}{l^2} = 0.$$

Theorem 2. *As* $N,\ k \longrightarrow \infty$ *in such a way that*

$$\frac{k}{N} = \gamma_N \leqslant 1 - \varepsilon \quad (\varepsilon > 0), \quad \frac{k^2}{N} \to \infty, \tag{7}$$

we have

$$\left.\begin{aligned} \mathbf{M}\nu_k &= N \ln \frac{1}{1-\gamma_N} + O(1), \\ \mathbf{D}\nu_k &= N\left(\frac{\gamma_N}{1-\gamma_N} - \ln \frac{1}{1-\gamma_N}\right) + O(1), \end{aligned}\right\} \tag{8}$$

$$m_3(l) = \mathbf{M}\,|\delta_l - \mathbf{M}\delta_l|^3 = O\left(\frac{l}{N}\right) \tag{9}$$

uniformly with respect to $l = 1, 2, \ldots, k$.

PROOF. The first formula in (8) follows immediately from (4) and (6). Integrating the inequalities

$$\frac{1}{x^2} \leqslant \frac{1}{[x]^2} \leqslant \frac{1}{(x-1)^2}$$

from $N-k+2$ to $N+1$, we get

$$\frac{k-1}{(N+1)(N-k+2)} \leqslant \sum_{l=N-k+2}^{N} \frac{1}{l^2} \leqslant \frac{k-1}{N(N-k+1)} \tag{10}$$

and, therefore,

$$\sum_{l=N-k+1}^{N} \frac{1}{l^2} = \frac{1}{N} \frac{\gamma_N}{1-\gamma_N} + O\left(\frac{1}{N^2}\right).$$

From this and (4) we arrive at the second of formulas (8). Let us prove (9).

Put $p_N = (l-1)/N$. Since

$$1 < \mathbf{M}\delta_l = \frac{1}{1-p_N} \leqslant \frac{1}{\varepsilon}, \qquad l = 1, \ldots, k,$$

we have

$$m_3(l) = \sum_{m=1}^{[\mathbf{M}\delta_l]} \left(\frac{1}{1-p_N} - m\right)^3 p_N^{m-1}(1-p_N) +$$
$$+ \sum_{m=[\mathbf{M}\delta_l]+1}^{\infty} \left(m - \frac{1}{1-p_N}\right)^3 p_N^{m-1}(1-p_N) \leqslant$$
$$\leqslant \mathbf{M}\delta_l \frac{p_N^3}{(1-p_N)^2} + \sum_{m=2}^{\infty} m^3 p_N^{m-1} \leqslant p_N \left(\frac{1}{\varepsilon^3} + \sum_{m=2}^{\infty} m^3 \varepsilon^{m-2}\right),$$

from which (9) follows, thus proving the theorem.

Theorem 3. *If* N, $k \longrightarrow \infty$ *in such a way that* $\frac{k^2}{2N} \to \lambda$ *(where* $0 < \lambda < \infty$*), then*

$$\mathbf{M}\nu_k = k + \lambda + O\left(\frac{1}{\sqrt{N}}\right), \qquad \mathbf{D}\nu_k = \lambda + O\left(\frac{1}{\sqrt{N}}\right).$$

PROOF. Let us write the difference $\nu_k - k$ in the form

$$\nu_k - k = (\delta_2 - 1) + \ldots + (\delta_k - 1). \tag{11}$$

From formulas (3) we see that

$$\mathbf{M}(\delta_l - 1) = \frac{\frac{l-1}{N}}{1 - \frac{l-1}{N}} = \frac{l-1}{N} + O\left(\frac{k^2}{N^2}\right),$$

$$\mathbf{D}(\delta_l - 1) = \left(\frac{1}{1 - \frac{l-1}{N}}\right)^2 - \frac{1}{1 - \frac{l-1}{N}} = \frac{l-1}{N} + O\left(\frac{k^2}{N^2}\right)$$

uniformly with respect to $l=2, \ldots, k$. If we sum the right-hand sides of these formulas with respect to l from 2 to k, the result in each case is equal to $\lambda + O(1/\sqrt{N})$, since $O(k^3/N^2) = O(1/\sqrt{N})$. From this and (11) the assertion of the theorem follows. Next let us find the limit distributions of ν_k.

Theorem 4. *Suppose that $k=N-c$, where $c \geqslant 0$ is a fixed integer. Then*

$$\lim_{N\to\infty} \mathbf{P}\left\{\frac{\nu_k}{N} - \ln N < x\right\} = \sum_{l=0}^{c} \frac{1}{l!}(e^{-x})^l \exp\{-e^{-x}\}.$$

PROOF. In equation (1.33)

$$\mathbf{P}\{\nu_k \leqslant n\} = \mathbf{P}\{\mu_0(n, N) \leqslant c\} \tag{12}$$

we set $n = [N(x + \ln N)]$. Then $\alpha = \frac{n}{N} = \ln N + x_N$, where $x_N \longrightarrow x$. Consequently, $n, N \longrightarrow \infty$ in the right-hand domain and, by virtue of Theorem 1.2, the distribution of $\mu_0(n, N)$ converges to a Poisson distribution with parameter $\lambda = e^{-x}$. Therefore, for the right-hand side of (12),

$$\mathbf{P}\{\mu_0(n, N) \leqslant c\} \to \sum_{l=0}^{c} \frac{\lambda^l}{l!} e^{-\lambda} \text{ as } N \longrightarrow \infty.$$

From this and (12), the assertion of the theorem follows.

We have taken advantage here of our limit theorem for μ_0. We shall prove for ν_k

Theorem 5. *Suppose that* $\frac{k}{N} = \gamma_N \to \gamma$ *(where* $0 \leqslant \gamma < 1$*) and* $\frac{k^2}{N} \to \infty$ *as* $N, k \longrightarrow \infty$. *Then,*

$$\lim_{N\to\infty} \mathbf{P}\left\{\frac{\nu_k - N \ln \frac{1}{1-\gamma_N}}{\sqrt{N\left(\frac{\gamma_N}{1-\gamma_N} - \ln\frac{1}{1-\gamma_N}\right)}} < x\right\} = \frac{1}{\sqrt{2\pi}} \int_{-\infty}^{x} e^{-u^2/2} du.$$

PROOF. Let us show that the hypotheses of Lyapunov's theorem are satisfied for the sum $\nu_k = \delta_1 + \ldots + \delta_k$. Using (9), we get

$$\sum_{l=1}^{k} \mathbf{M} |\delta_l - \mathbf{M}\delta_l|^3 = O(k\gamma_N)$$

and, therefore,

$$\frac{\left(\sum_{l=1}^{k} m_3(l)\right)^{1/3}}{\sqrt{\mathbf{D}\nu_k}} = \frac{O\left(k^{1/3}\gamma_N^{1/3}\right)}{N^{1/2}\left(\frac{\gamma_N}{1-\gamma_N} - \ln\frac{1}{1-\gamma_N}\right)^{1/2}}. \tag{13}$$

If $\gamma_N \longrightarrow \gamma$, where $0 < \gamma < 1$, the right-hand side of (13) is equal to $O(N^{-1/6})$. If $\gamma_N \longrightarrow 0$, we have

$$\frac{\gamma_N}{1-\gamma_N} - \ln\frac{1}{1-\gamma_N} \sim \frac{\gamma_N^2}{2} = \frac{1}{2}\left(\frac{k}{N}\right)^2$$

and, therefore, the right-hand side of (13) is equal to $O((k^2/N)^{-1/6})$. Thus, the hypotheses of Lyapunov's theorem are satisfied and the asymptotic normality of ν_k is proved.

Theorem 6. *If* $\frac{k^2}{2N} \to \lambda$, *where* $0<\lambda<\infty$, *as* $N, k \longrightarrow \infty$, *we have*

$$\mathbf{P}\{\nu_k - k = m\} \to \frac{\lambda^m}{m!} e^{-\lambda}.$$

PROOF. It follows from (1) and (1.32) that

$$\mathbf{M}x^{\nu_k} = \prod_{l=1}^{k} \left(1 - \frac{l-1}{N}\right) \frac{x}{1 - \frac{l-1}{N} x}.$$

Multiplying both sides of this equation by x^{-k}, we have

$$\varphi_N(x) = \mathbf{M}x^{\nu_k - k} = \prod_{l=1}^{k} \left(1 - \frac{l-1}{N}\right) \frac{1}{1 - \frac{l-1}{N} x},$$

from which we find that

$$\ln \varphi_N(x) = \sum_{l=1}^{k} \left[\ln\left(1 - \frac{l-1}{N}\right) - \ln\left(1 - \frac{l-1}{N} x\right)\right] =$$
$$= \frac{k^2}{2N}(x-1) + O\left(\frac{k^3}{N^2}\right) \to \lambda(x-1)$$

as $N \longrightarrow \infty$ and that

$$\lim_{N\to\infty} \varphi_N(x) = e^{\lambda(x-1)}.$$

The theorem is proved.

§3. Asymptotic Normality of the Number of Empty Cells

This section contains three proofs of the asymptotic normality of μ_0 in three overlapping domains:

1) the central domain;
2) the central domain and the left-hand intermediate domain;
3) the central domain and the right-hand intermediate domain.

Theorem 1. *In the central domain,*

$$\lim_{N\to\infty} \mathbf{P}\left\{\frac{\mu_0(n, N) - \mathbf{M}\mu_0}{\sqrt{\mathbf{D}\mu_0}} \leqslant x\right\} = \frac{1}{\sqrt{2\pi}} \int_{-\infty}^{x} e^{-u^2/2} du.$$

PROOF. By virtue of the asymptotic formulas (1.14), (1.15), we can prove for $\mathbf{M}\mu_0$, $\mathbf{D}\mu_0$ the asymptotic normality of the variable

$$\eta_N = \frac{\mu_0 - Ne^{-\alpha}}{\sigma\sqrt{N}}, \tag{1}$$

where $\alpha = n/N$ and $\sigma^2 = e^{-\alpha}(1-(1+\alpha)e^{-\alpha})$. If in (1.29) we set $x = x(t) = \exp\left(\frac{it}{\sigma\sqrt{N}}\right)$, the characteristic function

$$\Psi_N(t) = \mathbf{M}e^{it\eta_N} = \exp\left(-\frac{ite^{-\alpha}\sqrt{N}}{\sigma}\right)\mathbf{M}\exp\left(\frac{it\mu_0}{\sigma\sqrt{N}}\right)$$

can be expressed as

$$\Psi_N(t) = \exp\left(-\frac{ite^{-\alpha}\sqrt{N}}{\sigma}\right)\frac{1}{2\pi i}\frac{n!}{N^n}\oint (e^z + x(t) - 1)^N \frac{dz}{z^{n+1}}. \tag{2}$$

Let us transform (2) as follows. We replace $n!$ in accordance with Stirling's formula, setting $n = \alpha N$: $n! = \sqrt{2\pi\alpha N}(\alpha N)^{\alpha N}e^{-\alpha N}(1+o(1))$; we choose as the integration contour the circle $z = \alpha e^{iu}$, $-\pi \leqslant u \leqslant \pi$. Making the change of variable $v = u\sqrt{\alpha N}$, we obtain, as $N \to \infty$,

$$\Psi_N(t) \sim \frac{1}{\sqrt{2\pi}} \exp\left(-i\frac{t\sqrt{N}}{\sigma}e^{-\alpha}\right) \int_{-\pi\sqrt{\alpha N}}^{\pi\sqrt{\alpha N}} A^N(v)\, e^{-iv\sqrt{\alpha N}}\, dv, \quad (3)$$

where

$$A(v) = A_1(v) \cdot A_2(v),$$

$$A_1(v) = \exp\left[\alpha\left(e^{\frac{iv}{\sqrt{\alpha N}}} - 1\right)\right],$$

$$A_2(v) = 1 + \left(e^{\frac{it}{\sigma\sqrt{N}}} - 1\right) \exp\left(-\alpha e^{\frac{iv}{\sqrt{\alpha N}}}\right).$$

Suppose that t is fixed. Let us partition the integration domain in (3) into three parts:

$$S_1 = \{v : \delta\sqrt{\alpha N} \leqslant |v| \leqslant \pi\sqrt{\alpha N}\}, \quad 0 < \delta < 1,$$

$$S_2 = \left\{v : \sqrt[7]{N} \leqslant |v| \leqslant \delta\sqrt{\alpha N}\right\}, \quad S_3 = \left\{v : |v| \leqslant \sqrt[7]{N}\right\}.$$

In the domain S_1, as $N \longrightarrow \infty$

$$|A_1(v)| = \exp\left(-2\alpha \sin^2 \frac{v}{2\sqrt{\alpha N}}\right) \leqslant \exp\left(-2\alpha \sin^2 \frac{\delta}{2}\right) = q < 1, \quad (4)$$

since, by the hypotheses of the theorem, $\alpha \geqslant \alpha_0 > 0$. Furthermore,

$$|A_2(v)| \leqslant 1 + \frac{C}{\sqrt{N}}, \quad (5)$$

where $C > 0$ is a constant. Using (4) and (5), we see that

$$\left|\int_{S_1} A^N(v)\, e^{-iv\sqrt{\alpha N}}\, dv\right| \leqslant 2\pi\sqrt{\alpha N}\, q^N \left(1 + \frac{C}{\sqrt{N}}\right)^N \to 0 \quad (6)$$

as $N \longrightarrow \infty$. It can easily be verified that in the domain $S_2 \cup S_3$

$$\ln A_1(v) = -\frac{v^2}{2N} + iv\sqrt{\frac{\alpha}{N}} + O\left(\frac{|v|^3}{N^{3/2}}\right), \tag{7}$$

$$\ln A_2(v) = \frac{it}{\sigma\sqrt{N}} e^{-\alpha} + \frac{vt\sqrt{\alpha}}{\sigma N} e^{-\alpha} - \\ - \frac{t^2 e^{-\alpha}}{2N\sigma^2}(1 - e^{-\alpha}) + O\left(\frac{1+|v|^3}{N^{3/2}}\right).$$

These decompositions enable us to estimate $|A(v)|^N$ in the domain S_2:

$$|A(v)|^N < \exp\left\{-\frac{v^2}{2} + K_1\left(\frac{|v|^3}{\sqrt{N}} + |v| + 1\right)\right\}, \tag{8}$$

where $K_1 > 0$ is a constant. Since in S_2

$$\frac{|v|^3}{\sqrt{N}} = \frac{|v|}{\sqrt{\alpha N}} \sqrt{\alpha}\, v^2 < \delta \sqrt{\alpha}\, v^2, \quad |v| = \frac{v^2}{|v|} \leqslant \frac{v^2}{\sqrt[7]{N}},$$

we see from (8) that, if we choose δ sufficiently small, we have

$$|A(v)|^N < e^{-c_\delta v^2}, \quad c_\delta > 0.$$

Therefore,

$$\left|\int_{S_2} A^N(v)\, e^{-iv\sqrt{\alpha N}} dv\right| \leqslant 2 \int_{\sqrt[7]{N}}^{\delta\sqrt{\alpha N}} e^{-c_\delta v^2} dv \to 0 \tag{9}$$

as $N \longrightarrow \infty$.

The estimates (7) and (9) enable us to replace the integration domain in (3) by S_3. Hence

$$\Psi_N(t) \sim \frac{1}{\sqrt{2\pi}} \exp\left(-i\frac{t\sqrt{N}}{\sigma} e^{-\alpha}\right) \int_{-\sqrt[7]{N}}^{\sqrt[7]{N}} A^N(v)\, e^{-iv\sqrt{\alpha N}} dv \tag{10}$$

as $N \longrightarrow \infty$.

It follows from (7) that, in the domain $|v| \leqslant \sqrt[7]{N}$,

$$\ln\left[e^{-iv\sqrt{\alpha N}}\cdot A^{N}(v)\cdot\exp\left(-i\frac{t\sqrt{N}}{\sigma}e^{-\alpha}\right)\right]=$$

$$=-\frac{v^2}{2}+\frac{vt\sqrt{\alpha}}{\sigma}e^{-\alpha}-\frac{t^2}{2\sigma^2}(e^{-\alpha}-e^{-2\alpha})+O\left(\frac{1+|v|^3}{\sqrt{N}}\right)=$$

$$=-\frac{t^2}{2\sigma^2}(1-(1+\alpha)e^{-\alpha})\,e^{-\alpha}-\frac{1}{2}\left(v-\frac{t\sqrt{\alpha}}{\sigma}e^{-\alpha}\right)^2+O\left(N^{-1/14}\right).$$

Using this estimate in (10), we obtain

$$\Psi_N(t)\sim e^{-t^2/2}\frac{1}{\sqrt{2\pi}}\int\limits_{-\sqrt[7]{N}}^{\sqrt[7]{N}}\exp\left[-\frac{1}{2}\left(v-\frac{t\sqrt{\alpha}}{\sigma}e^{-\alpha}\right)^2\right]\times$$
$$\times\left(1+O\left(N^{-1/14}\right)\right)dv\to e^{-t^2/2},$$

from which the assertion of the theorem follows.

We shall give another proof of the asymptotic normality in the central domain and the right-hand intermediate domain.

Let us define

$$\varphi_{n,m}(t)=F_{m,N}(x)\,x^{-(N+n-m)e^{-n/N}},\qquad x=\exp\left(\frac{it}{\sigma\sqrt{N}}\right),$$
$$F_{m,N}(x)=\mathbf{M}x^{\mu_0(m,N)},\quad \sigma^2=e^{-n/N}\left[1-\left(1+\frac{n}{N}\right)e^{-n/N}\right]. \tag{11}$$

We note that the function $\varphi_{n,n}(t)$ coincides with the characteristic function $\Psi_N(t)$ of the random variable (1).

Lemma 1. *If* $n, N\longrightarrow\infty$ *in such a way that*

$$Ne^{-n/N}\to\infty,\quad n/N\geqslant\alpha_0>0,$$

we have

$$|\varphi_{n,m}(t)-\varphi_{n,n}(t)|=o\left(\frac{\ln^3 N}{\sqrt{N}}\right)$$

uniformly with respect to all m satisfying the inequality

$$|m-n| \leqslant \sqrt{n} \ln n.$$

PROOF. Let us multiply equation (1.5), namely,

$$\mathbf{P}\{\mu_0(m+1, N) = k\} = \left(1 - \frac{k}{N}\right)\mathbf{P}\{\mu_0(m, N) = k\} + \\ + \frac{k+1}{N}\mathbf{P}\{\mu_0(m, N) = k+1\},$$

by x^k and let us sum with respect to $k \geqslant 0$. After simple transformations, we have

$$F_{m+1,N}(x) - F_{m,N}(x) = \frac{1}{N}\left(\frac{1}{x} - 1\right)\sum_{k=0}^{\infty} kx^k \mathbf{P}\{\mu_0(m, N) = k\}. \tag{12}$$

Let us replace k in equation (12) with $(k - \mathbf{M}\mu_0) + \mathbf{M}\mu_0$. Since

$$\sum_{k=0}^{\infty} |k - \mathbf{M}\mu_0| \mathbf{P}\{\mu_0 = k\} \leqslant \sqrt{\mathbf{D}\mu_0},$$

we have from (12)

$$F_{m+1,N}(x) - F_{m,N}(x) = \\ = F_{m,N}(x)\left(\frac{1}{x} - 1\right)\frac{\mathbf{M}\mu_0}{N} + O\left(\frac{|x-1|\sqrt{\mathbf{D}\mu_0}}{N}\right).$$

Replacing $F_{m+1,N}$ and $F_{m,N}$ with $\varphi_{n,m+1}$ and $\varphi_{n,m}$ respectively, we find by formulas (11) that

$$\varphi_{n,m+1}(t) - \varphi_{n,m}(t) = \\ = \varphi_{n,m}(t)\left[x^{e^{-n/N}} - 1 + \left(\frac{1}{x} - 1\right)\frac{\mathbf{M}\mu_0}{N} x^{e^{-n/N}}\right] + \\ + O\left(\frac{|x-1|\sqrt{\mathbf{D}\mu_0}}{N}\right). \tag{13}$$

By the hypotheses of the lemma,

$$\frac{n}{N} = o(\ln N), \qquad \frac{m}{N} = \frac{n}{N} + o\left(\frac{\ln^{3/2} N}{\sqrt{N}}\right)$$

as $N \longrightarrow \infty$. Taking advantage of (1.14) and (1.15), we get

$$\frac{\mathbf{M}\mu_0(m, N)}{N} = e^{-n/N}\left(1 + o\left(\frac{\ln^{3/2} N}{\sqrt{N}}\right)\right), \quad \frac{\mathbf{D}\mu_0(m, N)}{N} = O(\sigma^2).$$

Furthermore,

$$\frac{1}{x} - 1 = -\frac{it}{\sigma\sqrt{N}} + O\left(\frac{1}{\sigma^2 N}\right), \quad |x - 1| = O\left(\frac{1}{\sigma\sqrt{N}}\right),$$

$$x^{e^{-n/N}} - 1 = \frac{it}{\sigma\sqrt{N}} e^{-n/N} + O\left(\frac{e^{-2n/N}}{\sigma^2 N}\right).$$

From this and (13) it follows that

$$\varphi_{n,m+1}(t) - \varphi_{n,m}(t) = o\left(\frac{\ln^{3/2} N}{N}\right).$$

Since

$$|\varphi_{n,m}(t) - \varphi_{n,n}(t)| \leqslant \sum_{l \in D_n} |\varphi_{n,l+1}(t) - \varphi_{n,l}(t)|, \qquad (14)$$

where $D_n = \{l : |l - n| < \sqrt{n} \ln n\}$, the left-hand side of (14) does not exceed $o\left(\frac{\ln^3 N}{\sqrt{N}}\right)$. The lemma is proved.

Let us define

$$K_{n,N}(t) = \sum_{m=0}^{\infty} \frac{n^m}{m!} e^{-n} \varphi_{n,m}(t), \qquad (15)$$

where $\varphi_{n,m}(t)$ is defined by (11).

Lemma 2. *If* $n, N \to \infty$ *in such a way that*

$$Ne^{-n/N} \to \infty, \qquad \frac{n}{N} \geqslant \alpha_0 > 0,$$

then

$$\lim_{N\to\infty} K_{n,N}(t) = e^{-t^2/2}.$$

PROOF. In the equation

$$\sum_{m=0}^{\infty} \frac{N^m}{m!} z^m \mathbf{M} x^{\mu_0(m,N)} = (e^z + x - 1)^N$$

let us express $F_{m,N}(x) = \mathbf{M}x^{\mu_0(m,N)}$ in terms of $\varphi_{n,m}(t)$ with the aid of (11). Taking

$$z = \frac{n}{N} x^{e^{-n/N}},$$

in the resulting equation and making some simplifications, we find that

$$K_{n,N}(t) = y^{-(N+n)} e^{n(y-1)} [1 + (x-1)\, e^{-ny/N}]^N, \quad (16)$$

where

$$x = \exp\left(\frac{it}{\sigma\sqrt{N}}\right), \qquad y = \exp\left(\frac{it}{\sigma\sqrt{N}} e^{-n/N}\right). \quad (17)$$

It is easy to verify that,

$$\begin{aligned} x - 1 &= \frac{it}{\sigma\sqrt{N}} - \frac{t^2}{2\sigma^2 N} + O\left(\frac{1}{(\sigma\sqrt{N})^3}\right), \\ y - 1 &= \frac{it}{\sigma\sqrt{N}} e^{-n/N} - \frac{t^2}{2\sigma^2 N} e^{-2n/N} + O\left(\frac{e^{-3n/N}}{(\sigma\sqrt{N})^3}\right) \end{aligned} \quad (18)$$

as $N \longrightarrow \infty$. Using (16)-(18) and (11), we see that

$$\ln K_{n,N}(t) = -\frac{t^2}{2} + O\left(\frac{1}{\sigma\sqrt{N}}\right)$$

as $N \longrightarrow \infty$.

By the hypotheses of the lemma, $\sigma\sqrt{N} \longrightarrow \infty$ as $N \longrightarrow \infty$. Therefore, the assertion of the lemma follows from the last equality.

Theorem 2. *In the central domain and in the right-hand intermediate domain, the random variable* $\mu_0(n, N)$ *is asymptotically normal with the parameters* $(\mathbf{M}\mu_0, \sqrt{\mathbf{D}\mu_0})$.

PROOF. We express $K_{n,N}(t)$ as the sum

$$K_{n,N}(t) = K_{n,N}^{(1)}(t) + K_{n,N}^{(2)}(t),$$

where $K_{n,N}^{(1)}(t)$ is the sum of those terms in the right-hand side of (15) corresponding to m such that

$$\left|\frac{m-n}{\sqrt{n}}\right| \leqslant \ln n.$$

As $n, N \longrightarrow \infty$,

$$|K_{n,N}^{(2)}(t)| \leqslant \sum_{|m-n|>\sqrt{n}\ln n} \frac{n^m}{m!}e^{-n} = \mathbf{P}\left\{\left|\frac{\xi_n - \mathbf{M}\xi_n}{\sqrt{\mathbf{D}\xi_n}}\right| > \ln n\right\} \rightarrow 0,$$

where ξ_n is a random variable, Poisson-distributed with parameter n.

By virtue of Lemma 1, we have

$$\varphi_{n,m}(t) = \varphi_{n,n}(t) + o\left(\frac{\ln^3 N}{\sqrt{N}}\right), \quad \frac{|m-n|}{\sqrt{n}} < \ln n.$$

Substituting this expression into $K_{n,N}^{(1)}(t)$, we get

$$K_{n,N}(t) = \varphi_{n,n}(t)\,\mathbf{P}\left\{\left|\frac{\xi_n - \mathbf{M}\xi_n}{\sqrt{\mathbf{D}\xi_n}}\right| < \ln n\right\} + o(1). \qquad (19)$$

Since, as $n, N \longrightarrow \infty$,

$$\mathbf{P}\left\{\left|\frac{\xi_n - \mathbf{M}\xi_n}{\sqrt{\mathbf{D}\xi_n}}\right| < \ln n\right\} \to 1$$

and (by virtue of Lemma 2)

$$K_{n,N}(t) \to e^{-t^2/2},$$

we have from (19) the result that

$$\lim_{N\to\infty} \varphi_{n,n}(t) = e^{-t^2/2},$$

thus proving the theorem.

It remains to prove the asymptotic normality of $\mu_0(n, N)$ in the left-hand intermediate domain. First we shall prove

Lemma 3. *The equation*

$$\frac{A(\varepsilon)}{1-x} = \exp\left(-\varepsilon\sqrt{\frac{x}{1-x} - \ln\frac{1}{1-x}}\right), \tag{20}$$

where $\varepsilon > 0$ *and* $0 < c \leqslant A(\varepsilon) < 1$, *has a unique root* $x(\varepsilon)$ *in the interval* $[0, 1)$. *Furthermore,*

$$x(\varepsilon) = 1 - A(\varepsilon) - \varepsilon A(\varepsilon)\sqrt{\frac{1-A(\varepsilon)}{A(\varepsilon)} + \ln A(\varepsilon)} + O(\varepsilon^2) \tag{21}$$

as $\varepsilon \longrightarrow 0$.

Proof. The function $\sqrt{x(1-x)^{-1} + \ln(1-x)}$ increases monotonically on the interval $[0, 1)$ from 0 to ∞ and, therefore, the right-hand side of (20) decreases monotonically from 1 to 0; the left-hand side of (20) increases from $A(\varepsilon) < 1$ to ∞ as x increases from 0 to 1. Therefore, there exists a unique solution $x(\varepsilon)$, $0 < x(\varepsilon) < 1 - c$. For $0 < x(\varepsilon) < 1 - c$, the right-hand side of (20) is equal to $1 + O(\varepsilon)$ as $\varepsilon \longrightarrow 0$. Therefore,

$$\frac{A(\varepsilon)}{1 - x(\varepsilon)} = 1 + O(\varepsilon)$$

and

$$x(\varepsilon)=1-A(\varepsilon)+O(\varepsilon).$$

Using this in (20), we obtain the approximation given by (21). The lemma is proved.

Theorem 3. *In the central domain and the left-hand intermediate domain, the random variable* $\mu_0(n, N)$ *is asymptotically normal with parameters* $(\mathbf{M}\mu_0, \sqrt{\mathbf{D}\mu_0})$.

PROOF. By hypotheses of the theorem, formulas (1.14) and (1.15) hold. Also,

$$\mathbf{D}\mu_0 \sim N\sigma^2 = Ne^{-\alpha}[1-(1+\alpha)e^{-\alpha}] \to \infty$$

as $N \to \infty$.

In the equation

$$\mathbf{P}\{\nu_k \leqslant n\} = \mathbf{P}\{\mu_0(n, N) \leqslant N-k\} \tag{22}$$

we set

$$n = N\ln\frac{1}{1-\gamma_N} + x\sqrt{N\left(\frac{\gamma_N}{1-\gamma_N} - \ln\frac{1}{1-\gamma_N}\right)} + O(1), \tag{23}$$

where $\gamma_N = k/N$. Then, by Theorem 2.5, for the left-hand side of (22), we see that

$$\mathbf{P}\{\nu_k \leqslant n\} = \mathbf{P}\left\{\frac{\nu_k - N\ln\frac{1}{1-\gamma_N}}{\sqrt{N\left(\frac{\gamma_N}{1-\gamma_N} - \ln\frac{1}{1-\gamma_N}\right)}} \leqslant x\right\} \to$$

$$\to \frac{1}{\sqrt{2\pi}}\int_{-\infty}^{x} e^{-u^2/2}du \tag{24}$$

as $N \to \infty$ provided the conditions of this theorem are satisfied:

$$N \to \infty, \quad \frac{k}{N} = \gamma_N \to \gamma \quad (0 \leqslant \gamma < 1), \quad \frac{k^2}{N} \to \infty. \qquad (25)$$

Let us rewrite (23) as follows:

$$\frac{e^{-\alpha}}{1-\gamma_N} = \exp\left(-\frac{x}{\sqrt{N}}\sqrt{\frac{\gamma_N}{1-\gamma_N} - \ln\frac{1}{1-\gamma_N}} + O\left(\frac{1}{N}\right)\right).$$

Taking advantage of Lemma 3, we have

$$\gamma_N = 1 - e^{-\alpha} - \frac{x}{\sqrt{N}}\sqrt{e^{-\alpha}[1-(1+\alpha)e^{-\alpha}]} + O\left(\frac{1}{N}\right),$$

or

$$k = N(1-e^{-\alpha}) - x\sqrt{N\sigma^2} + O(1), \qquad (26)$$

from which the relations (25) follow under the conditions stipulated in the theorem. Therefore, (24) is proved. Let us replace $N-k$ in (22) according to (26):

$$\mathbf{P}\{\mu_0(n, N) \leqslant N-k\} = \mathbf{P}\left\{\frac{\mu_0 - Ne^{-\alpha}}{\sigma\sqrt{N}} \leqslant x + O\left(\frac{1}{\sigma\sqrt{N}}\right)\right\}.$$

From this equation, (22), and (24) the assertion of the theorem follows.

§4. Poisson Limiting Distributions of the Number of Empty Cells

We proved in Section 1.1 the convergence of the distribution $\mu_0(n, N)$ to a Poisson distribution in the right-hand domain by the method of moments. This result was employed in proving Theorem 2.4 on the limiting behavior of the variable ν_k, where $N-k=c=\text{const}$, as $N \longrightarrow \infty$. Now we use Theorem 2.6 (which describes the behavior of ν_k when $N, k \longrightarrow \infty$ in such a way that $\frac{k^2}{2N} \to \lambda$) to investigate the behavior of $\mu_0(n, N)$ in the left-hand domain.

Theorem 1. *In the left-hand domain,*

$$\lim_{N\to\infty} \mathbf{P}\{\mu_0(n, N) - (N-n) = k\} = \frac{\lambda^k}{k!} e^{-\lambda}.$$

PROOF. In the right-hand side of (1.33)

$$\mathbf{P}\{\nu_k \leqslant n\} = \mathbf{P}\{\mu_0(n, N) \leqslant N-k\},$$

we set $k=n-m$; in the left-hand side, we set $n=k+m$, getting

$$\mathbf{P}\{\nu_k \leqslant k+m\} = \mathbf{P}\{\mu_0(n, N) - (N-n) \leqslant m\}. \tag{1}$$

In the left-hand domain, $\frac{n^2}{2N} = \lambda_N \to \lambda$ as $N \longrightarrow \infty$. From this and the fact that $k=n-m$, where $m=\text{const}$, we have

$$\frac{k^2}{2N} \to \lambda.$$

Thus, the conditions of Theorem 2.6 are satisfied. It follows from Theorem 2.6 that the left-hand side of (1) tends to

$$\sum_{l=0}^{m} \frac{\lambda^l}{l!} e^{-\lambda}.$$

From this and (1) the assertion of the theorem follows.

§5. Summary of Results on Limiting Distributions of the Number of Empty Cells

The results discussed in the previous sections describe fully all nonsingular limit distributions of $\mu_0(n, N)$. We have proved the following assertions for what happens as $n, N \longrightarrow \infty$:

1) in the left-hand domain,

$$\mathbf{M}\mu_0(n, N) \to \infty, \quad \mathbf{D}\mu_0(n, N) \to \lambda \ (\lambda > 0),$$
$$\mathbf{P}\{\mu_0(n, N) - (N - n) = k\} \to \frac{\lambda^k}{k!} e^{-\lambda};$$

2) in the left-hand, right-hand intermediate, or central domain,

$$\mathbf{M}\mu_0(n, N) \to \infty, \quad \mathbf{D}\mu_0(n, N) \to \infty,$$
$$\mathbf{P}\left\{\frac{\mu_0(n, N) - \mathbf{M}\mu_0(n, N)}{\sqrt{\mathbf{D}\mu_0(n, N)}} < x\right\} \to \frac{1}{\sqrt{2\pi}} \int_{-\infty}^{x} e^{-u^2/2} du;$$

3) in the right-hand domain,

$$\mathbf{M}\mu_0(n, N) \to \lambda, \quad \mathbf{D}\mu_0(n, N) \to \lambda,$$
$$\mathbf{P}\{\mu_0(n, N) = k\} \to \frac{\lambda^k}{k!} e^{-\lambda}.$$

If in the left-hand domain $\mathbf{D}\mu_0(n, N) \to \lambda = 0$, we have $\mathbf{P}\{\mu_0(n, N) = N - n\} \to 1$. In this case, there are few shots and they are distributed singly.

If in the right-hand domain $\mathbf{D}\mu_0(n, N) \to \lambda = 0$, there are many shots and $\mathbf{P}\{\mu_0(n, N) = 0\} \to 1$.

§6. Further Results. References

The scheme of allocation of particles discussed in this section can be found in the literature in many places. Theorem 1.2 on the convergence of the distribution of $\mu_0(n, N)$ to a Poisson distribution is due to von Mises [86]. This theorem is mentioned in many textbooks (see Bernstein [4], Feller [54], Rényi [92]). The Poisson law for the left-hand domain was proved by Békéssy [60]; we shall give another proof in Section 1.4. The asymptotic normality of $\mu_0(n, N)$ in the central domain was proved by Weiss [98], who used the method of moments. The asymptotic normality of $\mu_0(n, N)$ in intermediate domains was established by Rényi [91], whose methods we used in Section 1.3. Formulas (1.27) for the generating functions follow from the more general formulas of Sevast'yanov and Chistyakov [47].

Let us denote by $\nu_m(N, k)$ the number of trials after which k boxes first contain at least m shots each. The theorems on the limiting behavior of the variable $\nu_1(N, k)=\nu_k$ can be found in Section 1.2, where we used a part of Rényi's article [91]. The asymptotic behavior of $\nu_m(N, k)$ (where $m=\text{const}$) when N, $k \longrightarrow \infty$ in such a way that $N-k \leqslant c < \infty$ was studied by Erdös and Rényi [73] and Békéssy [61]. More refined asymptotic formulas for $\mathbf{M}\nu_m(N, N)$ and $\mathbf{D}\nu_m(N, N)$ were obtained by Brayton [65], Erdös and Rényi [73], and Newman and Shepp [88]. The asymptotic normality of $\nu_m(N, k)$(where $m=\text{const}$) as N, $k \longrightarrow \infty$ in such a way that $\frac{k}{N} \to \lambda$ (where $0<\lambda<1$) was proved by Békéssy [61]. The asymptotic behavior of $\nu_m(N, k)$ in the domains $\frac{k}{N} \to 0$ and $\frac{k}{N} \to 1$, $N-k \longrightarrow \infty$ and also as $N \longrightarrow \infty$ and $m \longrightarrow \infty$, was investigated in Ivchenko [20]. A thorough investigation of all limit distributions of $\nu_1(N, k)$ was carried out by Baum and Billingsley [59]. Sequential occupancy in the nonequiprobable case (see Ch. III) was studied by Barton and David [58] and Holst [82]. Similar investigation of the scheme of allocation of particles by complexes was carried out in Ivchenko and Medvedev [25].

Let us consider in more detail the problems of sequential occupancy in the following scheme.

Let

$$\xi_1, \xi_2, \ldots, \xi_n, \xi_{n+1}, \ldots, \xi_{n+s-1}, \ldots \tag{1}$$

be independent random variables with probability distributions

$$\mathbf{P}\{\xi_i=k\}=a_k,\ k=1,\ \ldots,\ m,\ i=1,\ 2,\ \ldots,\ n+s-1,\ \ldots\ .$$

The vectors

$$\boldsymbol{\xi}_i=(\xi_i,\ \xi_{i+1},\ \ldots,\ \xi_{i+s-1}).$$

are said to be s-chains.

Obviously, each s-chain can assume $N=m^s$ different values. The first $n+s-1$ terms of the sequence (1) contain n vectors $\boldsymbol{\xi}_i$. Following the conventional terminology, we may speak about allocation of vector-shots among $N=m^s$ boxes. However, the trials are now independent. This scheme is of some interest for investigating pseudorandom sequences used in Monte-Carlo simulations. Pseudorandom sequences can frequently be obtained on the basis of the recurrence relations

$$X_{i+s}=g(X_{i+s-1},\ \ldots,\ X_i)\bmod P, \tag{2}$$

where $i=1,\ 2,\ \ldots,$ the number s is fixed, the number P is an integer, and g is an integer-valued function (see Yermakov [17], Ch. II, §1). If the function g is complicated, investigation of the properties of the sequence (2) becomes rather difficult. In this case, the probabilistic approach can prove useful. We may consider the sequence $\{X_i\}$ to be random until first recurrence of the argument of the function g. The terms of this sequence will obviously recur from this instant of time on. Therefore, the time before the first recurrence in (2) can be estimated if we examine the time before the first recurrence of the s-chain in (1). The book by Sobol' [48] (Ch. I, §2) contains the theorem on the limiting distribution of the time before the first recurrence of the s-chain in (1) in the case $s=1$ It is also noted in this book that analogous theorems do not exist even for the case $s=2$. A considerably more general theorem was proved by Zubkov and Mikhaylov [18] in 1973. We shall state their results.

Let us define

$$\eta_{ij}=\begin{cases}1 & \text{if}\quad \boldsymbol{\xi}_i=\boldsymbol{\xi}_j,\\ 0 & \text{if}\quad \boldsymbol{\xi}_i\neq\boldsymbol{\xi}_j,\end{cases}\qquad \varepsilon_{ij}=\begin{cases}0 & \text{if}\quad \boldsymbol{\xi}_i=\boldsymbol{\xi}_j,\\ 1 & \text{if}\quad \boldsymbol{\xi}_i\neq\boldsymbol{\xi}_j.\end{cases}$$

If the vectors ξ_i and ξ_j coincide, then the vectors ξ_{i+1} and ξ_{j+1} will coincide if the variables ξ_{i+s} and ξ_{j+s} coincide. Hence, for $|i-j|>s$,

$$\mathbf{P}\{\xi_{i+1}=\xi_{j+1}|\xi_i=\xi_j\}=\mathbf{P}\{\xi_{i+s}=\xi_{j+s}\}=\sum_{k=1}^{m}a_k^2=A\,.$$

(If $|i-j|\leqslant s$, we do not in general have equality.) In other words, the coincidence leads to a run of coincidences the length of which has an asymptotically geometric distribution with generating function

$$\frac{z(1-A)}{1-Az}. \tag{3}$$

Let us define

$$\nu(s,n)=\sum_{1\leqslant i<j\leqslant n}\eta_{ij}\varepsilon_{i-1,j-1},\qquad \zeta(s,n)=\sum_{1\leqslant i<j\leqslant n}\eta_{ij},$$

$$\tau_s=\min\{n:\zeta(s,n)>0\}.$$

The variable $\nu(s,n)$ is equal to the number of recurrent runs; $\zeta(s,n)$ is the number of recurrences in $n+s-1$ trials; τ_s is the number of trials before the first recurrence. We have the following theorems:

Theorem 1. *If* s, m, $a_1,\ldots,a_m$ *are such that, as* $n\longrightarrow\infty$,

$$\frac{n(n-1)}{2}A^s(1-A)\to\lambda>0,\qquad s^t n\left(\max_{1<i\leqslant m}a_i\right)^s\to 0$$

for any t, *then*

1) $\mathbf{M}z^{\nu(s,n)}\to e^{\lambda(z-1)}$, $\quad\mathbf{P}\{\nu(s,n)=k\}\to\frac{\lambda^k}{k!}e^{-\lambda}$.

2) *if* $A\longrightarrow\rho\in[0,1)$, *we have*

$$\mathbf{M}z^{\zeta(s,n)} \to \exp\left\{\lambda \frac{(1-\rho)z}{1-\rho z} - 1\right\}. \tag{4}$$

Theorem 2. *If* $s, m, a_1, \ldots, a_m$ *are such that, as* $n \longrightarrow \infty$,

$$A^s \to 0, \qquad s^t \left(\frac{\max a_i}{\sqrt{A}}\right)^s \to 0$$

for any t, *then*

$$\mathbf{P}\left\{\frac{\tau_s}{\sqrt{A^s(1-A)}} \leqslant x\right\} \to 1 - e^{-x^2/2}.$$

The limiting distribution given by (4) is the distribution of the variable

$$\zeta(s, n) = \theta_1 + \theta_2 + \ldots + \theta_{\nu(s,n)},$$

where the θ_i are independent with common generating function given by (3), and the number of terms $\nu(s, n)$ is distributed according to a Poisson law.

Theorems 1 and 2 were proved by Zubkov and Mikhaylov [18]. They also obtained in [18] the limit distribution of the variables $\max\{s: \zeta(s, n) > 0\}$. Theorem 3.2.1 of Sevast'yanov [46] was used in the proof. Different generalizations of the results obtained in [18] are due to Mikhaylov ([37], [38], [39]).

The properties of the sequence (1) with $a_k = 1/m$ were studied in Belyayev [3]; Belyayev also investigated in [2], [1] the probabilities of failure of a given number of s-chains in the case in which the sequence (1) is a Markov chain (either simple or compound). Some results on the number of nonappearing states in Markov chains of a specific form were obtained by Kolchin and Chistyakov [33].

Chapter II

EQUIPROBABLE ALLOCATIONS

§1. Generating Functions. Moments

In this chapter, we shall examine the scheme of equiprobable allocations in which n particles are allocated in N cells independently; the probability of any one particle's getting into any one cell is equal to $1/N$. Let us denote by $\mu_r(n, N)$ the random variable equal to the number of cells containing exactly r particles, $r=0, 1, 2, \ldots$. We shall sometimes write $\mu_r(n)$ or μ_r instead of $\mu_r(n, N)$.

Chapter I dealt with asymptotic properties of the distribution of μ_0, the number of empty cells. The present chapter discusses the properties of the distribution of μ_r as $n, N \longrightarrow \infty$.

The possible values m_r of the random variables μ_r are related by

$$\sum_{r=0}^{n} m_r = N, \qquad \sum_{r=0}^{n} r m_r = n. \tag{1}$$

The first relation in (1) yields the total number of cells; the second yields the total number of particles. If the m_r satisfy (1), the joint distribution of μ_r, $r=0, 1, \ldots, n$ depends on the probabilities

$$\mathbf{P}\{\mu_r = m_r,\ r = 0, 1, \ldots, n\} = \frac{N!n!}{N^n \prod_{r=0}^{n} [(r!)^{m_r} \cdot m_r!]}. \tag{2}$$

For all remaining values of m_r, the probabilities $\mathbf{P}\{\mu_r=m_r, r=0, \ldots, n\}$ are zero.

We can obtain formula (2) as follows. Set $\xi_{ij}=1$ if the ith particle gets into the jth cell (according to some fixed numbering scheme) and $\xi_{ij}=0$ in the remaining cases. Then each realization of the allocation is associated with a (0, 1)-matrix $\mathrm{H}=\|\xi_{ij}\|$ of order $n\times N$ such that $\sum_{j=1}^{N} \xi_{ij}=1$ for $1\leqslant i\leqslant n$. The number of these matrices is N^n and they are equiprobable. Let us denote by $\eta_j = \sum_{i=1}^{n} \xi_{ij}$ the number of particles in the jth cell.

Let us count the number of matrices H with given $\{m_r\}$. We can choose columns of a given composition from among N columns in $\frac{N!}{m_0!m_1!\ldots m_n!}$ ways. Each such choice involves

$$\frac{n!}{\prod_{r=0}^{n} (r!)^{m_r}}$$

arrangements of rows. Similarly, $(2!)^{m_2}(3!)^{m_3}\ldots(n!)^{m_n}$ permutations of rows do not change the matrix H. From this formula (2) follows.

Formula (2) is not very useful in studying distribution laws either of the individual μ_r or $\mu_{r_1}, \ldots, \mu_{r_s}$ jointly. There are, however, more interesting methods.

Let us first calculate the first and second moments of μ_r. Let us introduce indicators θ_{ri}, taking θ_{ri} equal to 1 if there are r

particles in the ith cell and equal to zero otherwise. Then $\mu_r = \sum_{i=1}^{N} \theta_{ri}$ and $\mathbf{M}\mu_r = N\mathbf{P}\{\theta_{ri}=1\}$. Since

$$\mathbf{P}\{\theta_{ri} = 1\} = C_n^r \frac{1}{N^r}\left(1 - \frac{1}{N}\right)^{n-r},$$

we have

$$\mathbf{M}\mu_r = NC_n^r \frac{1}{N^r}\left(1 - \frac{1}{N}\right)^{n-r}. \tag{3}$$

The second moments of μ_r can be found in a similar way. If $r \neq t$, then

$$\mathbf{M}\mu_r\mu_t = \mathbf{M}\sum_{i,j=1}^{N} \theta_{ri}\theta_{tj} = \sum_{i,j=1}^{N} \mathbf{M}\theta_{ri}\theta_{tj} =$$
$$= \sum_{i\neq j} \mathbf{P}\{\theta_{ri} = \theta_{tj} = 1\} = N(N-1)\frac{n^{[r+t]}}{r!t!N^{r+t}}\left(1 - \frac{2}{N}\right)^{n-r-t}, \tag{4}$$

since $\theta_{ri}\theta_{ti} = 0$. Noting that $\theta_{ri}^2 = \theta_{ri}$, we have

$$\mathbf{M}\mu_r^2 = \mathbf{M}\sum_{i,j=1}^{N} \theta_{ri}\theta_{rj} = \sum_{i=1}^{N} \mathbf{M}\theta_{ri} + \sum_{i\neq j} \mathbf{M}\theta_{ri}\theta_{rj} =$$
$$= \mathbf{M}\mu_r + N(N-1)\frac{n^{[2r]}}{(r!)^2 N^{2r}}\left(1 - \frac{2}{N}\right)^{n-2r}. \tag{5}$$

We can obtain the asymptotic behavior of $\mathbf{M}\mu_r$ and $\mathbf{Cov}(\mu_r, \mu_t)$ from (3)-(5). We shall use the notation $\alpha = n/N$, $p_r(\alpha) = \frac{\alpha^r}{r!}e^{-\alpha}$. Also, we shall sometimes write p_r instead of $p_r(\alpha)$.

Theorem 1. *For any n, N, and r, we have the inequality*

$$\mathbf{M}\mu_r \leqslant Np_r(\alpha)e^{r/N}. \tag{6}$$

If $\alpha = o(N)$ *as* $n, N \to \infty$, *we have for fixed* r *and* t

$$\mathbf{M}\mu_r = Np_r(\alpha) + p_r(\alpha)\left(r - \frac{\alpha}{2} - \frac{C_r^2}{\alpha}\right) + O\left(\frac{1}{N}\right), \tag{7}$$

$$\mathbf{Cov}(\mu_r, \mu_t) \sim N\sigma_{rt}(\alpha), \tag{8}$$

where

$$\begin{aligned} \sigma_{rr}(\alpha) &= p_r(\alpha)\left(1 - p_r(\alpha) - p_r(\alpha)\,\frac{(\alpha - r)^2}{\alpha}\right), \\ \sigma_{rt}(\alpha) &= -\,p_r(\alpha)\,p_t(\alpha)\left(1 + \frac{(\alpha - r)(\alpha - t)}{\alpha}\right). \end{aligned} \tag{9}$$

PROOF. Inequality (6) follows from (3) and the estimates $C_n^r \leqslant \frac{n^r}{r!}$, $1 - \frac{1}{N} \leqslant e^{-1/N}$. The asymptotic formula (7) holds for $r \geqslant 2$ if in (3) we set $C_n^r = \frac{1}{r!}\left(n^r - C_r^2 n^{r-1} + O(n^{r-2})\right)$:

$$\begin{aligned} \mathbf{M}\mu_r &= N\,\frac{n^r - C_r^2 n^{r-1} + O(n^{r-2})}{r!N^r}\exp\left\{(n-r)\ln\left(1 - \frac{1}{N}\right)\right\} = \\ &= \frac{N}{r!}\left(\alpha^r - C_r^2\frac{\alpha^r}{N\alpha} + O\left(\frac{\alpha^{r-2}}{N^2}\right)\right) \times \\ &\qquad \times \exp\left\{-(n-r)\left(\frac{1}{N} + \frac{1}{2N^2} + O\left(\frac{1}{N^3}\right)\right)\right\} = \\ &= \frac{N}{r!}\alpha^r\left(1 - \frac{C_r^2}{N\alpha} + O\left(\frac{1}{N^2\alpha^2}\right)\right)\exp\left(-\alpha - \frac{\alpha}{2N} + \frac{r}{N}\right) \times \\ &\qquad \times \left(1 + O\left(\frac{1+\alpha}{N^2}\right)\right) = \frac{N}{r!}\alpha^r e^{-\alpha}\left(1 - \frac{C_r^2}{N\alpha} + O\left(\frac{1}{N^2\alpha^2}\right)\right) \times \\ &\qquad \times \left(1 - \frac{\alpha}{2N} + \frac{r}{N}\right)\left(1 + O\left(\frac{1+\alpha^2}{N^2}\right)\right), \end{aligned}$$

from which we get (7). For $r < 2$, the remainder term $O(1/N^2\alpha^2)$ is zero. The asymptotic estimate given in (8) can be obtained in a similar way from (4), (5), (7), noting that $\mathbf{Cov}(\mu_r, \mu_t) = \mathbf{M}\mu_r\mu_t - \mathbf{M}\mu_r\mathbf{M}\mu_t$.

We shall use the following terminology from now on. If n, $N \longrightarrow \infty$ in such a way that

$$0 < c_1 \leqslant \alpha = \frac{n}{N} \leqslant c_2 < \infty, \tag{10}$$

where c_1 and c_2 are constants, we say that $n, N \rightarrow \infty$ *in the central domain.* It follows from Theorem 1 that in this case all $\mathbf{M}\mu_r$ and $\mathbf{D}\mu_r$ increase asymptotically in proportion to N, and we see that, as $\alpha \longrightarrow \alpha_0$,

$$\lim_{n,N\to\infty} \frac{\mathbf{D}\mu_r}{\mathbf{M}\mu_r} = 1 - p_r(\alpha_0) - p_r(\alpha_0)\frac{(\alpha_0 - r)^2}{\alpha_0} < 1.$$

Consider now μ_r with $r \geqslant 2$. We shall call the domain of variation n, $N \longrightarrow \infty$, in which

$$\alpha \to 0, \quad 0 < \lim \mathbf{M}\mu_r = \lambda < \infty, \tag{11}$$

the *left-hand r-domain.* It is easily seen that $\lim \mathbf{D}\mu_r = \lambda$ in this domain. It follows from (7) that, in this case,

$$\lambda = \lim \mathbf{M}\mu_r = \lim N \frac{\alpha^r}{r!} e^{-\alpha} = \lim \frac{n^r}{r! N^{r-1}}, \tag{12}$$

i.e., $n \sim (r!\lambda/N)^{1/r} N$. Denote $\bar{\mu}_r = \sum_{k>r} \mu_k$. Let us estimate $\mathbf{M}\bar{\mu}_r$ in the left-hand r-domain using (6):

$$\mathbf{M}\bar{\mu}_r = \sum_{k=r+1}^{n} \mathbf{M}\mu_k \leqslant N \sum_{k=r+1}^{n} \frac{\alpha^k}{k!} e^{-\alpha + \frac{k}{N}} \leqslant$$

$$\leqslant N \frac{\alpha^r}{r!} \sum_{k=1}^{\infty} \alpha^k = O(\alpha) \to 0.$$

By Chebyshev's inequality, we then obtain in the left-hand r-domain

$$\mathbf{P}\{\mu_r > 0\} \leqslant \mathbf{M}\mu_r \to 0. \tag{13}$$

In particular, it follows that in the left-hand 2-domain each μ_r for $r > 2$ is zero with probability 1 in the limit. Also, from equations (1),

$$\mu_0 + \mu_1 + \mu_2 = N, \qquad \mu_1 + 2\mu_2 = n,$$

we have

$$\mu_0 = N - n + \mu_2, \qquad \mu_1 = n - 2\mu_2, \tag{14}$$

that is, for

$$0 < \frac{n^2}{2N} \to \lambda < \infty, \tag{15}$$

the asymptotic properties of the random variables

$$\mu_2, \quad \mu_0 - N + n, \quad \frac{n - \mu_1}{2} \tag{16}$$

are all the same. Hence, from now on we shall assume that the left-hand r-domain, for $r = 0, 1, 2$, is given by (15).

We define the *left-hand intermediate r-domain for* $r \geqslant 2$ as the n, $N \longrightarrow \infty$ for which

$$\alpha \to 0, \quad \lim \mathbf{M}\mu_r = \lim N p_r(\alpha) = \lim \frac{n^r}{r!N^{r-1}} = \infty. \tag{17}$$

In this case,

$$\lim_{n, N \to \infty} \frac{\mathbf{D}\mu_r}{\mathbf{M}\mu_r} = 1. \tag{18}$$

The *left-hand intermediate r domain for* $r=0$, 1, 2 is defined by the relations

$$\frac{n}{N} \to 0, \qquad \frac{n^2}{2N} \to \infty. \tag{19}$$

Let r be any nonnegative integer. The domain $n, N \longrightarrow \infty$ in which

$$\alpha \to \infty, \quad \mathbf{M}\mu_r \to \lambda, \quad 0<\lambda<\infty, \tag{20}$$

is called the *right-hand r-domain.* In this domain, we also have $\mathbf{D}\mu_r \to \lambda$.

It follows from (20) that $N \frac{\alpha^r}{r!} e^{-\alpha} \to \lambda$, or

$$\alpha = r \ln \alpha + \ln N - \ln (r!\, \lambda) + o(1), \tag{21}$$

from which we have $\alpha = (1+o(1)) \ln N$. Substituting this expression into (21), we see that, in the right-hand r-domain,

$$\alpha = \ln N + r \ln \ln N - \ln (r! \lambda) + o(1). \tag{22}$$

We shall refer to the domain in which

$$\alpha \to \infty, \quad \mathbf{M}\mu_r \to \infty \tag{23}$$

as the *right-hand intermediate r-domain.* It follows from (23) that in this domain

$$\alpha \longrightarrow \infty, \qquad \alpha - \ln N - r \ln \ln N \longrightarrow -\infty \tag{24}$$

and $\mathbf{D}\mu_r/\mathbf{M}\mu_r \to 1$. We define $\underline{\mu}_r = \sum_{k=0}^{r-1} \mu_k$.

As seen from (22), we have in the right-hand r-domain

$$\mathbf{M}\underline{\mu}_r \to 0. \tag{25}$$

From this it follows by Chebyshev's inequality that $\mathbf{P}\{\mu_r=0\}\to 1$.

Thus, we can give the following picture of the asymptotic behavior of μ_r as $n, N\to\infty$ with α varying from 0 to ∞. For very small α, i.e., for $N\alpha^2\longrightarrow 0$, with probability tending to 1, $\mu_0=N-n$, $\mu_1=n$, and all the remaining $\mu_r=0$. Next, in the left-hand 2-domain with nonzero probability $\mu_2>0$ and $\mathbf{M}\mu_2\to\text{const}$, and all the remaining $\mu_r=0$ for $r=3, 4, \ldots$. If we get into the left-hand r-domain, μ_r will be positive with nonzero probability, $\mathbf{M}\mu_r\to\text{const}$, $\mu_s\xrightarrow{\mathbf{P}}\infty$ for all $s<r$, and $\mathbf{P}\{\mu_t=0\}\longrightarrow 1$ for all $t>r$. In the central domain, for any fixed r we have $\mu_r\longrightarrow\infty$ in probability. In the right-hand 0-domain, only μ_0 first becomes finite and nonzero. In the right-hand r-domain $\mathbf{P}\{\mu_0=\ldots=\mu_{r-1}=0\}\to 1$, $\mathbf{M}\mu_r\to\text{const}$ and $\mu_r>0$ with nonzero probability, and all the remaining $\mu_t\xrightarrow{\mathbf{P}}\infty$ for any fixed $t>r$.

To study the limiting distribution laws for μ_r, we shall use generating functions. Let $r_1<r_2<\ldots<r_s$ be fixed. Denote

$$F_{n,r_1\ldots r_s}(x_1,\ldots,x_s)=\\ =\sum_{k_1,\ldots,k_s}\mathbf{P}\{\mu_{r_1}=k_1,\ldots,\mu_{r_s}=k_s\}\,x_1^{k_1}\ldots x_s^{k_s} \tag{26}$$

and

$$\Phi^{(N)}_{r_1\ldots r_s}(z;\,x_1,\ldots,x_s)=\sum_{n=0}^{\infty}\frac{N^n z^n}{n!}F_{n,r_1\ldots r_s}(x_1,\ldots,x_s). \tag{27}$$

Theorem 2. *The generating function given by* (27) *is as follows:*

$$\Phi^{(N)}_{r_1\ldots r_N}(z;\,x_1,\ldots,x_s)=\\ =\left[e^z+\frac{z^{r_1}}{r_1!}(x_1-1)+\ldots+\frac{z^{r_s}}{r_s!}(x_s-1)\right]^N. \tag{28}$$

PROOF. Divide N cells into two groups with N_1 cells in one and $N-N_1=N_2$ cells in the other. Since, with probability $C_n^{n_1}\frac{N_1^{n_1}N_2^{n_2}}{N^n}$, n_1 particles get into the first group and $n_2=n-n_1$ particles get into the second group, we have from the formula for the joint probability:

$$F_{n,r_1\ldots r_s}(x_1, \ldots, x_s) = \sum_{n_1+n_2=n} \frac{n!N_1^{n_1}N_2^{n_2}}{n_1!n_2!N^n} F_{n_1,r_1\ldots r_s} F_{n_2,r_1\ldots r_s},$$

from which, by the definition given in (27), we get:

$$\Phi^{(N)}_{r_1\ldots r_s} = \Phi^{(N_1)}_{r_1\ldots r_s} \cdot \Phi^{(N_2)}_{r_1\ldots r_s}. \tag{29}$$

It follows from (29) that $\Phi^{(N)}_{r_1\ldots r_s}=\left[\Phi^{(1)}_{r_1\ldots r_s}\right]^N$. Therefore, to prove (28), it suffices to prove that

$$\Phi^{(1)}_{r_1\ldots r_s}(z; x_1, \ldots, x_s) =$$
$$= e^z + \frac{z^{r_1}}{r_1!}(x_1-1) + \ldots + \frac{z^{r_s}}{r_s!}(x_s-1).$$

This last formula follows from the fact that $\mathbf{P}\{\mu_r(n, 1)=0\}=1$ for $n\neq r$ and $\mathbf{P}\{\mu_r(r, 1)=1\}=1$.

Let us introduce also the generating function

$$\Phi^{(N)}_{r_1\ldots r_s}(z) = \sum_{n=0}^{\infty} \frac{z^n N^n}{n!} \mathbf{P}\{\mu_{r_1}(n, N) = \ldots = \mu_{r_s}(n, N) = 0\}.$$

Since $\Phi^{(N)}_{r_1\ldots r_s}(z) = \Phi^{(N)}_{r_1\ldots r_s}(z; 0, \ldots, 0)$, from Theorem 2 we get

Corollary 1.

$$\Phi^{(N)}_{r_1\ldots r_s}(z) = \left(e^z - \frac{z^{r_1}}{r_1!} - \ldots - \frac{z^{r_s}}{r_s!}\right)^N. \tag{30}$$

Setting $s=1$ in (28) and (30), we arrive at

Corollary 2.

$$\Phi_r^{(N)}(z;\ x) = \left[e^z + \frac{z^r}{r!}(x-1)\right]^N, \tag{31}$$

$$\Phi_r^{(N)}(z) = \left(e^z - \frac{z^r}{r!}\right)^N. \tag{32}$$

Using (31) and (32), we can express the probabilities $\mathbf{P}\{\mu_r(n,\ N)=k\}$ in terms of the probabilities $\mathbf{P}\{\mu_r=0\}$.

Corollary 3.

$$\mathbf{P}\{\mu_r(n,\ N)=k\} =$$
$$= C_N^k \frac{n^{[rk]}}{(r!)^k N^{rk}} \left(1-\frac{k}{N}\right)^{n-rk} \mathbf{P}\{\mu_r(n-rk,\ N-k)=0\}. \tag{33}$$

PROOF. Differentiating (31) k times with respect to x and then setting x equal to zero, we get

$$N^{[k]}\left(\frac{z^r}{r!}\right)^k \left(e^z - \frac{z^r}{r!}\right)^{N-k}. \tag{34}$$

The coefficient of x^k in (31) is equal to the function (34) divided by $k!$. To obtain the probability $\mathbf{P}\{\mu_r(n,\ N)=k\}$, we need next to multiply the coefficient of z^n in the expansion of the function (34) by $\frac{n!}{N^n k!}$. The coefficient of z^{n-kr} in the expansion of the function $\left(e^z - \frac{z^r}{r!}\right)^{N-k}$ is equal to

$$\frac{(N-k)^{n-kr}}{(n-kr)!}\ \mathbf{P}\{\mu_r(n-kr,\ N-k)=0\},$$

from which we obtain (33).

The multivariate generalization of (33) is obtained in a similar way.

Corollary 4.

$$\mathbf{P}\{\mu_{r_i}=k_i,\ i=1,\ldots,s\}=$$

$$=\frac{N^{[k]}\cdot n^{[(k,r)]}}{\prod\limits_{i=1}^{s}\left(k_i!\,(r_i!)^{k_i}\right)}\ \frac{1}{N^{(k,r)}}\left(1-\frac{k}{N}\right)^{n-(k,r)}\times$$

$$\times\mathbf{P}\{\mu_{r_i}(n-(k,r),N-k)=0,\ i=1,\ldots,s\}, \qquad (35)$$

where $k=k_1+\ldots+k_s$ *and* $(k,r)=k_1r_1+\ldots+k_sr_s$.

Finally, using the generating function (28) we can obtain an expression for any factorial moment:

Theorem 3. *For any choice of nonnegative integers* k_i,

$$\mathbf{M}\mu_{r_1}^{[k_1]}\ldots\mu_{r_s}^{[k_s]}=N^{[k]}\frac{n^{[(k,r)]}}{N^{(k,r)}(r_1!)^{k_1}\ldots(r_s!)^{k_s}}\left(1-\frac{k}{N}\right)^{n-(r,k)}, \qquad (36)$$

where $k=k_1+\ldots+k_s$ *and* $(k,r)=\sum\limits_{i=1}^{s}k_ir_i$.

PROOF. Differentiating (28), we obtain

$$\left.\frac{\partial^k\Phi}{\partial x_1^{k_1}\ldots\partial x_s^{k_s}}\right|_{x_1=\ldots=x_s=1}=N^{[k]}\left(\frac{z^{r_1}}{r_1!}\right)^{k_1}\ldots\left(\frac{z^{r_s}}{r_s!}\right)^{k_s}e^{z(N-k)}. \qquad (37)$$

The coefficient z^n in the expansion of the left and right sides of (37) multiplied by $n!/N^n$ yields (36). The theorem is proved.

In particular, for $s=1$ we have:

Corollary 5. *The factorial moments of* μ_r *are given by the formula*

$$\mathbf{M}\mu_r^{[k]}=N^{[k]}\frac{n^{[rk]}}{(r!)^k}\left(1-\frac{k}{N}\right)^{n-rk}. \qquad (38)$$

§2. Asymptotic Normality in the Central Domain

By Theorem 1.1, the covariance $\mathbf{Cov}(\mu_r,\mu_t)$ increases in proportion to N for any r, t in the central domain. Therefore, in this case, for fixed $0 \leqslant r_1 < r_2 < \ldots < r_s$, the random vector

$$\frac{\mu_{r_1} - \mathbf{M}\mu_{r_1}}{\sqrt{N}}, \quad \frac{\mu_{r_2} - \mathbf{M}\mu_{r_2}}{\sqrt{N}}, \quad \ldots, \frac{\mu_{r_s} - \mathbf{M}\mu_{r_s}}{\sqrt{N}} \tag{1}$$

has the limiting covariance matrix

$$\mathbf{B} = \|\sigma_{r_i r_j}(\alpha)\|, \quad i, j = 1,\ldots,s, \tag{2}$$

where the $\sigma_{rt}(\alpha)$ are defined by (1.9). We denote by B^2 the determinant of the covariance matrix $\mathbf{B}$. For brevity, we write $\Delta_r = \alpha - r$. Then

$$B^2 = \left| x_i \delta_{ij} - p_{r_i r_j} \left(1 + \frac{\Delta_{r_i} \Delta_{r_j}}{\alpha}\right) \right|,$$

where $x_i = p_{r_i}$, $i = 1, \ldots, s$. To compute B^2, we expand it with respect to the "variables" $x_1, x_2, \ldots, x_s$ and their products. The coefficients in this expansion are diagonal minors of the form

$$\left| -p_{r_i} p_{r_j} \left(1 + \frac{\Delta_{r_i} \Delta_{r_j}}{\alpha}\right) \right| \tag{3}$$

of different orders. The first-order determinant is equal to $-p_{r_i}^2 \left(1 + \frac{\Delta_{r_i}^2}{\alpha}\right)$; the second-order determinant is equal to $p_{r_i}^2 p_{r_j}^2 \frac{(\Delta_{r_i} - \Delta_{r_j})^2}{\alpha}$; as can easily be verified, all determinants of higher orders are zero. Hence, we infer that

$$B^2 = p_{r_1} \ldots p_{r_s} \left[1 - \sum_{k=1}^{s} p_{r_k} \left(1 + \frac{\Delta_{r_k}^2}{\alpha}\right) + \right.$$
$$\left. + \frac{1}{2} \sum_{k,l=1}^{s} \frac{p_{r_k} p_{r_l}}{\alpha} (\Delta_{r_k} - \Delta_{r_l})^2 \right]. \tag{4}$$

If we introduce the notation $q=1-p_{r_1}-p_{r_2}-\cdots-p_{r_s}$, we can express (4) as:

$$B^2=\frac{p_{r_1}\cdots p_{r_s}q}{\alpha}\left[\alpha-\sum_{k=1}^{s}p_{r_k}\Delta_{r_k}^2-\frac{1}{q}\left(\sum_{k=1}^{s}p_{r_k}\Delta_{r_k}\right)^2\right]. \tag{5}$$

Lemma 1. *For any* $0<\alpha<\infty$ *we have* $B^2>0$.

PROOF. By virtue of (5), the inequality $B^2>0$ is equivalent to the inequality

$$q\left(\alpha-\sum_{k=1}^{s}p_{r_k}\Delta_{r_k}^2\right)>\left(\sum_{k=1}^{s}p_{r_k}\Delta_{r_k}\right)^2. \tag{6}$$

Since $\sum_{r=0}^{\infty}p_r=1$, $\sum_{r=0}^{\infty}rp_r=\alpha$, $\sum_{r=0}^{\infty}p_r\Delta_r=0$, $\sum_{r=0}^{\infty}p_r\Delta_r^2=\alpha$, inequality (6) can be written

$$\sum_{r\notin S}^{\infty}p_r\sum_{r\notin S}p_r\Delta_r^2>\left(\sum_{r\notin S}p_r\Delta_r\right)^2, \tag{7}$$

where $S=(r_1,\ldots,r_s)$. Inequality (7) is in fact the Cauchy-Bunyakovskiy inequality. Equality is impossible here since for equality to hold it is necessary and sufficient that Δ_r, where $r\notin S$, be independent of r and, in our case, $\Delta_r=\alpha-r$. The lemma is proved.

Since B^2 is a continuous function of α, we get from Lemma 1

Corollary 1. *If* $0<\alpha_0\leqslant\alpha_1<\infty$, *then*

$$\min_{\alpha_0\leqslant\alpha\leqslant\alpha_1}B^2>0.$$

Further on, we shall need the cofactors of the elements $\sigma_{r_kr_l}$ of the matrix (2) expressed in the forms

$$B_{kk} = \frac{p_{r_1} \dots p_{r_s}(q + p_{r_k})}{\alpha p_{r_k}} \left[\alpha - \sum_{i \neq k} p_{r_i} \Delta_{r_i}^2 - \right.$$

$$\left. - \frac{1}{q + p_{r_k}} \left(\sum_{i \neq k} p_{r_i} \Delta_{r_i} \right)^2 \right], \quad (8)$$

$$B_{kl} = p_{r_1} \dots p_{r_s} \left[1 + \frac{\Delta_{r_k} \Delta_{r_l}}{\alpha} - \right.$$

$$\left. - \sum_{i=1}^{s} p_{r_i} \frac{(\Delta_{r_i} - \Delta_{r_k})(\Delta_{r_i} - \Delta_{r_l})}{\alpha} \right], \quad k \neq l. \quad (9)$$

The cofactor B_{kk} is computed in the same way as B^2 is. To compute B_{kl}, where $k \neq l$, we proceed as follows: Suppose that $k < l$. The cofactor B_{kl} is equal to the determinant of the matrix (2) multiplied by $(-1)^{k+l}$ after the kth row and the lth column are removed. To compute this determinant, we move the kth column $l-k-1$ places to the right. This operation yields the factor $(-1)^{l-k-1}$. The determinant thus obtained will be of the same structure as that of the determinant B^2. In this case, the $(l-1)$st row consists of elements of the form $-p_{r_i} p_{r_l}\left(1 + \frac{\Delta_{r_i} \Delta_{r_l}}{\alpha}\right)$, and the $(l-1)$st column consists of elements of the form $-p_{r_i} p_{r_k}\left(1 + \frac{\Delta_{r_i} \Delta_{r_k}}{\alpha}\right)$. By the method used in computing B^2, we arrive at (9).

Formulas (8) and (9) can be rewritten

$$B_{kk} = \frac{B^2}{p_{r_k}} + \frac{B^2}{q} + \frac{p_{r_1} \dots p_{r_s}}{\alpha q} \left(\Delta_{r_k} q + \sum_{i=1}^{s} p_{r_i} \Delta_{r_i} \right)^2, \quad (10)$$

$$B_{kl} = \frac{B^2}{q} + \frac{p_{r_1} \dots p_{r_s}}{\alpha q} \left(\Delta_{r_k} q + \sum_{i=1}^{s} p_{r_i} \Delta_{r_i} \right) \times$$

$$\times \left(\Delta_{r_l} q + \sum_{i=1}^{s} p_{r_i} \Delta_{r_i} \right). \quad (11)$$

For n, $N \longrightarrow \infty$ in the central domain, we obtain the local normal theorem for $\mu_{r_1}, \mu_{r_2}, \ldots, \mu_{r_s}$.

Formula (1.35) can be written as the product

$$\mathbf{P}\{\mu_{r_i}=k_i, \quad i=1, 2, \ldots, s\} = X_1 \cdot X_2 \cdot X_3, \tag{12}$$

where

$$X_1 = \frac{N!}{(N-k)!k_1! \ldots k_s!}, \qquad X_2 = \frac{n!}{(r_1!)^{k_1} \ldots (r_s!)^{k_s} N^n}, \tag{13}$$

$$X_3 = \frac{(N')^{n'}}{n'!} \mathbf{P}\{\mu_{r_i}(n', N') = 0, \quad i = 1, 2, \ldots, s\},$$

$$N' = N - k, \ n' = n - (k, r), \quad k = \sum_{i=1}^{s} k_i, \ (k, r) = \sum_{i=1}^{s} k_i r_i.$$

Asymptotic formulas for X_1 and X_2 as $N \longrightarrow \infty$ are easily obtained by using Stirling's formula.

The factor X_3 is the coefficient of $z^{n'}$ in (1.30); hence we infer that

$$X_3 = \frac{1}{2\pi i} \int\limits_{|z|=\text{const}} \frac{1}{z} \left[\frac{A(z)}{z^{\gamma}}\right]^{\lambda} dz, \tag{14}$$

where $\lambda = N'$, $\gamma = n'/N'$,

$$A(z) = e^z - \frac{z^{r_1}}{r_1!} - \ldots - \frac{z^{r_s}}{r_s!}. \tag{15}$$

An asymptotic formula for the integral in (14) can be obtained by the saddle-point method. Let us define in the right-hand plane the function $f(z) = \ln A(z) - \gamma \ln z$ that is real on the real axis. As integration contour let us take the circle $|z| = x_\gamma$, where x_γ is the real root of the equation $f'(x) = 0$. Let us first establish the existence of this x_γ. Express (15) as $A(z) = \sum_{k=0}^{\infty} a_k z^k$. Here, $a_k \geqslant 0$ and $a_k > 0$ for $k > r_s$.

Lemma 2. *If* $a_0 = a_1 = \ldots = a_{n_1-1} = 0$, $a_{n_1} > 0$, *the equation* $f'(x) = 0$ *has a unique real positive root* x_γ *for any* $\gamma > n_1$

PROOF. The equation $f'(x) = 0$ for $x > 0$ is equivalent to the equation

$$xA'(x) - \gamma A(x) = \sum_{k=0}^{\infty} (k - \gamma) a_k x^k =$$
$$= - \sum_{k=0}^{[\gamma]} (\gamma - k) a_k x^k + \sum_{k=[\gamma]+1}^{\infty} (k - \gamma) a_k x^k = 0,$$

or to the equation

$$G_\gamma(x) = - \sum_{k=n_1}^{[\gamma]} (\gamma - k) a_k x^{k-[\gamma]} +$$
$$+ \sum_{k=[\gamma]+1}^{\infty} (k - \gamma) a_k x^{k-[\gamma]} = 0.$$

It is easily seen that, for $0 < x < x'$, we have the inequality $G_\gamma(x) < G_\gamma(x')$. Furthermore, $G_\gamma(x) \longrightarrow \infty$ as $x \longrightarrow \infty$, and if $\gamma > n_1$, then $G_\gamma(x) < 0$ for small x. The lemma follows.

The next lemma establishes the asymptotic properties of the integral in (14):

Lemma 3. *Suppose that for some* $\gamma \in [\gamma_0, \gamma_1]$, *where* $0 < \gamma_0 \leqslant \gamma_1 < \infty$ *are constants, the unique root of the equation* $f(x) = 0$ *is* x_γ *and that there exists a* $\delta > 0$ *such that*

$$f''(x_\gamma) \geqslant \delta, \quad x_\gamma \geqslant \delta \quad (\gamma \in [\gamma_0, \gamma_1]).$$

Then

$$I_\lambda = \frac{1}{2\pi i} \int_{|z|=x_\gamma} \frac{1}{z} \left[\frac{A(z)}{z^\gamma} \right]^\lambda dz =$$
$$= \frac{1}{x_\gamma \sqrt{2\pi f''(x_\gamma) \lambda}} e^{\lambda f(x_\gamma)} \left[1 + O\left(\frac{1}{\sqrt{\lambda}} \right) \right] \tag{16}$$

uniformly with respect to $\gamma \in [\gamma_0, \gamma_1]$ *as* $\lambda \longrightarrow \infty$.

PROOF. Let us represent I_λ as the sum $I_1 + I_2$, where

$$I_1 = \frac{1}{2\pi} \int_{-\varepsilon}^{\varepsilon} e^{\lambda f(z)} d\theta,$$

$$I_2 = \frac{1}{2\pi} \int_{\varepsilon < |\theta| \leqslant \pi} \left[\frac{A(z)}{z^\gamma} \right]^\lambda d\theta, \quad z = x_\gamma e^{i\theta}, \quad \varepsilon > 0. \tag{17}$$

Since $a_h \geqslant 0$, we have everywhere on the circle $|z| = x_\gamma$ except at $z \neq x_\gamma$

$$\left| \frac{A(z)}{z^\gamma} \right| < \frac{A(x_\gamma)}{x_\gamma^\gamma}.$$

We estimate the variable I_2 as follows:

$$|I_2| \leqslant \frac{1}{2\pi} \left[\frac{A(x_\gamma)}{x_\gamma^\gamma} \right]^\lambda \int_{\varepsilon < |\theta| \leqslant \pi} |\Psi_\gamma(\theta)|^\lambda \, d\theta, \tag{18}$$

where $\Psi_\gamma(\theta) = A(x_\gamma e^{i\theta}) / A(x_\gamma)$ for any $\gamma \in [\gamma_0, \gamma_1]$ is the characteristic function of the discrete distribution with maximal step equal to one. The function $\Psi_\gamma(\theta)$ depends continuously on (γ, θ) for $\gamma_0 \leqslant \gamma \leqslant \gamma_1$, $0 \leqslant |\theta| \leqslant \pi$. Let us show that

$$q = \sup_{\substack{\gamma \in [\gamma_0, \gamma_1] \\ \varepsilon \leqslant |\theta| \leqslant \pi}} |\Psi_\gamma(\theta)| < 1. \tag{19}$$

Let us choose a sequence $(\gamma_k, \theta_k) \to (\tilde{\gamma}, \tilde{\theta})$ as $k \to \infty$, such that $\lim_{k \to \infty} |\Psi_{\gamma_k}(\theta_k)| = q$. Next we choose from the sequence γ_k a subsequence $\gamma_{\tilde{k}}$ such that $\Psi_{\gamma_{\tilde{k}}}(\theta)$ has the limit $\Psi_{\tilde{\gamma}}(\theta)$. Then $q = |\Psi_{\tilde{\gamma}}(\tilde{\theta})|$. Since the coefficients in the expansion of $\Psi_{\gamma_{\tilde{k}}}(\theta)$ in powers of $e^{i\theta}$ do not approach zero as $k \to \infty$, the maximal step of the distribution corresponding to $\Psi_{\tilde{\gamma}}(\theta)$ is equal to one.

Therefore,

$$q = |\Psi_{\widetilde{\gamma}}(\widetilde{\theta})| < 1 \quad \text{if} \quad 0 < \varepsilon \leqslant |\theta| \leqslant \pi.$$

We obtain from the estimates (18) and (19)

$$I_2 = O\left(\left[q\,\frac{A(x_\gamma)}{x_\gamma^\gamma}\right]^\lambda\right) \tag{20}$$

Since $f(x_\gamma e^{i\theta}) = u(\theta) + iv(\theta)$, where $u(-\theta) = u(\theta)$ and $v(-\theta) = -v(\theta)$, the integral I_1 can be written as

$$I_1 = \frac{1}{\pi} e^{\lambda u(0)} \int_0^\varepsilon \cos \lambda v(\theta)\, e^{-\lambda[u(0)-u(\theta)]} d\theta. \tag{21}$$

The function $f(z)$ is analytic in a neighborhood of the point $z = x_\gamma$. Since $f'(x_\gamma) = 0$, we infer that

$$f(z) = f(x_\gamma) + \frac{f''(x_\gamma)}{2}(z - x_\gamma)^2 + \dots .$$

Hence, for $z = x_\gamma e^{i\theta}$,

$$f(x_\gamma e^{i\theta}) = f(x_\gamma) - \frac{f''(x_\gamma)}{2} x_\gamma^2 \theta^2 + \dots .$$

It is not difficult to obtain the following expansions and estimates (uniform for $|\theta| < \varepsilon$, $\gamma \in [\gamma_0, \gamma_1]$):

$$u(\theta) = u(0) - \frac{x_\gamma^2}{2} f''(x_\gamma)\,\theta^2 + O(\theta^3), \qquad v(\theta) = O(\theta^3). \tag{22}$$

It follows from (22) that

$$\cos \lambda v(\theta) = 1 + O(\lambda\theta^3). \tag{23}$$

Using (22) and (23) and setting $\theta = \frac{1}{x_\gamma}\sqrt{\frac{2y}{\lambda f''(x_\gamma)}}$ in (21) we obtain $I_1 = I_{11} + I_{12}$, where

$$I_{11} = K(\lambda) \int_0^{\sqrt[4]{\lambda}} y^{-1/2} \left(1 + O\left(\frac{y^{3/2}}{\sqrt{\lambda}}\right)\right) \exp\left(-y + O\left(\frac{y^{3/2}}{\sqrt{\lambda}}\right)\right) dy,$$

$$I_{12} = K(\lambda) \int_{\sqrt[4]{\lambda}}^{\varepsilon_1 \lambda} y^{-1/2} \left(1 + O\left(\frac{y^{3/2}}{\sqrt{\lambda}}\right)\right) \exp\left(-y + O\left(\frac{y^{3/2}}{\sqrt{\lambda}}\right)\right) dy,$$

$$K(\lambda) = \frac{e^{\lambda u(0)}}{x_\nu \pi \sqrt{2\lambda f''(x_\nu)}}, \qquad \varepsilon_1 = \frac{1}{2} \varepsilon^2 f''(x_\nu)\, x_\nu^2.$$

Since $\frac{y^{3/2}}{\sqrt{\lambda}} = y \sqrt{\frac{y}{\lambda}} < y \sqrt{\varepsilon_1}$, it follows that, for sufficiently small ε_1 and $0 < y < \varepsilon_1 \lambda$,

$$\exp\left(-y + O\left(\frac{y^{3/2}}{\sqrt{\lambda}}\right)\right) = O(e^{-y/2}), \quad y^{-1/2} O\left(\frac{y^{3/2}}{\sqrt{\lambda}}\right) = O(\sqrt{y}).$$

Therefore,

$$I_{12} = O\left[K(\lambda) \int_{\sqrt[4]{\lambda}}^{\infty} \sqrt{y}\, e^{-y/2} dy\right] = O\left(\frac{K(\lambda)}{\sqrt{\lambda}}\right). \tag{24}$$

In the integral I_{11}, we have $O\left(\frac{y^{3/2}}{\sqrt{\lambda}}\right) \to 0$ uniformly in y. Therefore,

$$I_{11} = K(\lambda) \int_0^{\sqrt[4]{\lambda}} y^{-1/2} \left(1 + O\left(\frac{y^{3/2}}{\sqrt{\lambda}}\right)\right) e^{-y} dy =$$

$$= K(\lambda) \int_0^{\infty} y^{-1/2} e^{-y} dy + O\left(\frac{K(\lambda)}{\sqrt{\lambda}}\right).$$

From this and formulas (20) and (24), using the well-known formula $\sqrt{\pi}=\Gamma\left(\frac{1}{2}\right)=\int\limits_0^\infty y^{-1/2}e^{-y}dy$, we get the assertion of the lemma.

We shall now formulate the main theorem:

Theorem 1. *In the central domain as* $n, N \longrightarrow \infty$, *we have uniformly with respect to* $\alpha = \frac{n}{N} \in [\alpha_0, \alpha_1]$, $0 < \alpha_0 \leqslant \alpha_1 < \infty$, *and* $k_i = Np_{r_i} + u_i\sqrt{N}$, $|u_i| \leqslant C < \infty$,

$$\mathbf{P}\{\mu_{r_i} = k_i,\ i = 1, \ldots, s\} =$$

$$= \frac{1}{(2\pi N)^{s/2}\, B} \exp\left(-\frac{1}{2B^2}\sum_{i,j=1}^{s} B_{ij}u_iu_j\right)\left(1 + O\left(\frac{1}{\sqrt{N}}\right)\right), \qquad (25)$$

where $p_r = \frac{\alpha^r}{r!}e^{-\alpha}$ *and* B^2 *and* B_{ij} *are defined by* (5), (8), *and* (9).

Proof. In (13), let us set $n = \alpha N$, $k_i = Np_{r_i} + u_i\sqrt{N}$. Then,

$$n' = N[\alpha - (r, p)] - (r, u)\sqrt{N}, \qquad N' = Nq - u\sqrt{N},$$

where

$$u = \sum_{i=1}^{s} u_i,\ (r, p) = \sum_{i=1}^{s} r_i p_{r_i},\ (r, u) = \sum_{i=1}^{s} r_i u_i,\ q = 1 - \sum_{i=1}^{s} p_{r_i}.$$

Using Stirling's formula and substituting $r_i!$ for $\alpha^{r_i}e^{-\alpha}p_{r_i}^{-1}$, we obtain, as $N \longrightarrow \infty$,

$$\ln X_1 = -\ln(2\pi N)^{s/2} - \ln\sqrt{p_{r_1}\ldots p_{r_s}q} -$$

$$-N\left(\sum_{i=1}^{s} p_{r_i}\ln p_{r_i} + q\ln q\right) + \sqrt{N}\left(u\ln q - \sum_{i=1}^{s} u_i \ln p_{r_i}\right) -$$

$$-\frac{1}{2}\left(\sum_{i=1}^{s}\frac{u_i^2}{p_{r_i}} + \frac{u^2}{q}\right) + O\left(\frac{1}{\sqrt{N}}\right), \qquad (26)$$

$$\ln X_2 = \ln \sqrt{2\pi\alpha N} + \\ + N\left[\alpha \ln \alpha - (p, r) \ln \alpha - \alpha q + \sum_{i=1}^{s} p_{r_i} \ln p_{r_i}\right] - \\ - \sqrt{N}\left[(r, u) \ln \alpha - \alpha u - \sum_{i=1}^{s} u_i \ln p_{r_i}\right] + O\left(\frac{1}{\sqrt{N}}\right). \quad (27)$$

Let us estimate X_3. By Lemma 2, the equation $\frac{\partial f(x, \gamma)}{\partial x} = 0$, where

$$f(x, \gamma) = \ln A(x) - \gamma \ln x, \quad (28)$$

has a positive root x_γ. Then $z = x_\gamma$ is a saddle-point for (14). The variables x_γ and γ are related by the equation

$$\gamma = \frac{1}{\bar{q}}\left(x_\gamma - (r, \bar{p})\right), \quad (29)$$

where

$$(r, \bar{p}) = \sum_{i=1}^{s} r_i \bar{p}_{r_i}, \quad \bar{q} = 1 - \sum_{i=1}^{s} \bar{p}_{r_i}, \quad \bar{p}_r = p_r(x_\gamma) = \frac{x_\gamma^r}{r!} e^{-x_\gamma}.$$

In particular, for $x_\gamma = \alpha$ we write $\gamma = \beta$, where

$$\beta = \frac{\alpha - (r, p)}{q}. \quad (30)$$

Let us show that

$$\left.\frac{\partial^2 f(x, \gamma)}{\partial x^2}\right|_{\substack{x=\alpha \\ \gamma=\beta}} = \frac{B^2}{\alpha q^2 p_{r_1} \cdots p_{r_s}}. \quad (31)$$

To this end, we differentiate (28) twice with respect to x and set $x = \alpha$, $\gamma = \beta$:

$$\left.\frac{\partial^2 f}{\partial x^2}\right|_{\substack{x=\alpha\\ \gamma=\beta}} = \frac{1}{q^2}\left\{\left[1-\sum_{i=1}^{s}\frac{r_i(r_i-1)}{\alpha^2}p_{r_i}\right]q - \right.$$

$$\left. -\left(1-\sum_{i=1}^{s}\frac{r_i}{\alpha}p_{r_i}\right)^2\right\}+\frac{\beta}{\alpha^2}.$$

Substituting $\beta=(\alpha-(r,\ p))/q$ and $r_i=\alpha-\Delta_i$, we obtain

$$\alpha q^2\left.\frac{\partial^2 f}{\partial x^2}\right|_{\substack{x=\alpha\\ \gamma=\beta}} =$$

$$= q\left[1-\sum_{i=1}^{s}p_{r_i}\frac{\Delta_i^2}{\alpha}-\frac{1}{\alpha q}\left(\sum_{i=1}^{s}p_{r_i}\Delta_i\right)^2\right]=\frac{B^2}{p_{r_1}\cdots p_{r_s}},$$

from which (31) follows.

For large N, we can expand

$$\gamma=\frac{n'}{N'}=\frac{n-(k,r)}{N-k}=\frac{N[\alpha-(r,p)]-(r,u)\sqrt{N}}{Nq-u\sqrt{N}}$$

in powers of $t=N^{-1/2}$ and use the notation in (30):

$$\gamma=\beta+\frac{t}{q}[\beta u-(r,u)]+t^2\frac{u}{q^2}[\beta u-(r,u)]+O(t^3). \tag{32}$$

Equation (29) with the left side (32) has the root $x_\gamma=\alpha$ for $t=0$ by virtue of (30). By the implicit-function theorem, for any fixed α, equation (29) defines x_γ as a function of the variable t in a neighborhood of the point $t=0$. It is easy to obtain the expansion

$$x_\gamma=\alpha+t\frac{p_{r_1}p_{r_2}\cdots p_{r_s}q}{B}[\beta u-(r,u)]+O(t^2), \tag{33}$$

where the estimate $O(t^2)$ is uniform over $\alpha\in[\alpha_0,\ \alpha_1]$, $0<\alpha_0\leqslant\alpha_1<\infty$.

To estimate the integral X_3 in (14), we use Lemma 3. Let us show that the conditions of this lemma are satisfied. It is seen from (33) that $x_\gamma \sim \alpha$ as $t \longrightarrow 0$; since $0 < \alpha_0 \leqslant \alpha \leqslant \alpha_1 < \infty$, there exist α_0 and α_1 such that $0 < \alpha_0 \leqslant x_\gamma \leqslant \alpha_1 < \infty$ for large N. By virtue of (32), (33) and Corollary 1, we infer that

$$\min_{\alpha_0 \leqslant \alpha \leqslant \alpha_1} \left. \frac{\partial^2 f(x, \gamma)}{\partial x^2} \right|_{x=x_\gamma} \geqslant \delta > 0$$

for large N. Therefore, the conditions of Lemma 3 are satisfied with $\lambda = N'$ and

$$\ln X_3 = -\ln x_\gamma \sqrt{2\pi N' \left. \frac{\partial^2 f(x, \gamma)}{\partial^2 x} \right|_{x=x_\gamma}} + N' f(x_\gamma, \gamma) + O\left(\frac{1}{\sqrt{N'}}\right). \tag{34}$$

We expand $f(x_\gamma, \gamma)$ in powers of $t = N^{-1/2}$:

$$f(x_\gamma, \gamma) = f(\alpha, \beta) + \left. \frac{\partial f}{\partial t} \right|_{t=0} \frac{1}{\sqrt{N}} + \left. \frac{\partial^2 f}{\partial t^2} \right|_{t=0} \frac{1}{2N} + O\left(\frac{1}{N^{3/2}}\right), \tag{35}$$

where

$$\frac{\partial f}{\partial t} = \frac{\partial f}{\partial x_\gamma} \cdot \frac{dx_\gamma}{dt} + \frac{\partial f}{\partial \gamma} \cdot \frac{\partial \gamma}{\partial t} = \frac{\partial f}{\partial \gamma} \cdot \frac{\partial \gamma}{\partial t} = -\frac{\partial \gamma}{\partial t} \ln x_\gamma$$

(since $\partial f / \partial x_\gamma = 0$ is identically zero for all t by virtue of (29)) and

$$\frac{\partial^2 f}{\partial t^2} = -\frac{1}{x_\gamma} \frac{\partial \gamma}{\partial t} \frac{\partial x_\gamma}{\partial t} - \frac{\partial^2 \gamma}{\partial t^2} \ln x_\gamma.$$

Using (32) and (33), we obtain

$$\left. \frac{\partial f}{\partial t} \right|_{t=0} = -\frac{1}{q} [\beta u - (r, u)] \ln \alpha, \tag{36}$$

$$\left. \frac{\partial^2 f}{\partial t^2} \right|_{t=0} = -\frac{p_{r_1} \cdots p_{r_s}}{\alpha B^2} [\beta u - (r, u)]^2 - \frac{2u [\beta u - (r, u)]}{q^2} \ln \alpha. \tag{37}$$

From (35)-(37) and the equations

$$f(\alpha, \beta) = \alpha + \ln q - \beta \ln \alpha, \; N' = Nq - u\sqrt{N},$$

we have

$$\ln X_3 = N(\alpha q + q \ln q - \beta q \ln \alpha) - \\ - \sqrt{N}[\alpha u - (r, u) \ln \alpha + u \ln q] - \\ - \frac{p_{r_1} \cdots p_{r_s} q}{2\alpha B^2} [\beta u - (r, u)^2] - \ln \sqrt{2\pi N \frac{B^2 \alpha}{q p_{r_1} \cdots p_{r_s}}} + O\left(\frac{1}{\sqrt{N}}\right). \tag{38}$$

Equation (25) follows from (26), (27), (30), (38), (12), (13), (5), (10), and (11). The theorem is proved.

For $r_1 = r$ and $s = 1$, a univariate local theorem follows from Theorem 1:

Theorem 2. *In the central domain, as* $n, N \longrightarrow \infty$, *we have, uniformly with respect to* $\alpha = \frac{n}{N} \in [\alpha_0, \alpha_1]$, $0 < \alpha_0 \leqslant \alpha_1 < \infty$, *and* $k = Np_r(\alpha) + u\sqrt{N}$, $|u| \leqslant C < \infty$,

$$\mathbf{P}\{\mu_r = k\} = \frac{1}{\sqrt{2\pi N \sigma_{rr}(\alpha)}} e^{-\frac{u^2}{2\sigma_{rr}(\alpha)}} \left(1 + O\left(\frac{1}{\sqrt{N}}\right)\right), \tag{39}$$

where $\sigma_{rr}(\alpha)$ *is defined by* (1.9).

We can obtain an integral theorem for variables $\mu_{r_1}, \ldots, \mu_{r_s}$ from the local theorem in the usual way. Let us state the multivariate integral theorem as well as the univariate integral theorem.

Theorem 3. *In the central domain, as* $n, N \longrightarrow \infty$, *we have, uniformly with respect to* $\alpha = \frac{n}{N} \in [\alpha_0, \alpha_1]$, *where* $0 < \alpha_0 \leqslant \alpha_1 < \infty$,

$$\mathbf{P}\left\{\left(\frac{\mu_{r_i}-Np_{r_i}}{\sqrt{N}},\ i=1,\ldots,s\right)\in G\right\}=$$

$$=\frac{1}{(2\pi)^{s/2}B}\int\limits_G\cdots\int\exp\left\{-\frac{1}{2B^2}\sum_{i,j=1}^{s}B_{ij}u_iu_j\right\}du_1\ldots du_s+o(1),\tag{40}$$

where G is a measurable domain.

Theorem 4. *In the central domain as* $n, N\longrightarrow\infty$, *we have uniformly with respect to* $\alpha=\frac{n}{N}\in[\alpha_0,\alpha_1], 0<\alpha_0\leqslant\alpha_1<\infty$,

$$\mathbf{P}\left\{\frac{\mu_r-Np_r}{\sqrt{N}}<x\right\}=\frac{1}{\sqrt{2\pi\sigma_{rr}(\alpha)}}\int\limits_{-\infty}^{x}e^{-\frac{u^2}{2\sigma_{rr}(\alpha)}}du+o(1).\tag{41}$$

§3. Limit distributions of μ_r for $r\geqslant 2$

In the previous section we proved asymptotic normality of the random variables μ_r in the central domain, where $n, N\longrightarrow\infty$ and $0<\alpha_0\leqslant\alpha\leqslant\alpha_1<\infty$.

In the present section, we shall show that, as $n, N\longrightarrow\infty$, the univariate distributions of μ_r, for $r\geqslant 2$, approach a normal distribution in a wider domain where $Np_r\longrightarrow\infty$ and approach a Poisson distribution if $\alpha\longrightarrow 0$ or if $\alpha\longrightarrow\infty$.

Let η_i denote the number of particles in the ith cell, for $i=1, 2,\ldots, N$, in the case of equiprobable allocation of n particles in N cells. The random variables $\eta_1,\ldots,\eta_N$ have polynomial distributions. To study random variables with polynomial distributions, we can use the following relationship between a polynomial distribution and a Poisson distribution: if $\xi_1,\ldots,\xi_N$ are independent identically distributed random variables with a Poisson distribution with parameter α, then

$$\mathbf{P}\{\eta_i=k_i,\ i=1,\ldots,N\}=$$
$$=\mathbf{P}\{\xi_i=k_i,\ i=1,\ \ldots,\ N\,|\,\xi_1+\ldots+\xi_N=n\},\tag{1}$$

where $k_1, \ldots, k_N$ are nonnegative integers such that $k_1+\ldots+k_N=n$. This representation is easy to verify and is useful in investigating limit distributions of random variables $\mu_r(n, N)$ and those of an order sequence $\eta_{(1)}, \ldots, \eta_{(N)}$ obtained as a result of rearranging $\eta_1, \ldots, \eta_N$ in nondecreasing order.

Denote by $\xi_1^{(r)}, \ldots, \xi_N^{(r)}$ the independent identically distributed random variables whose distributions are related to the distributions of $\xi_1, \ldots, \xi_N$ as follows:

$$\mathbf{P}\{\xi_i^{(r)}=k\}=\mathbf{P}\{\xi_i=k \mid \xi_i \neq r\}.$$

We introduce the notation:

$$p_r=\frac{\alpha^r e^{-\alpha}}{r!}, \quad \zeta_N=\xi_1+\ldots+\xi_N, \qquad \zeta_N^{(r)}=\xi_1^{(r)}+\ldots+\xi_N^{(r)}.$$

The following lemma reduces study of the asymptotic behavior of the random variables $\mu_r(n, N)$ to proof of local theorems for independent summands.

Lemma 1.

$$\mathbf{P}\{\mu_r(n, N)=k\}=C_N^k p_r^k(1-p_r)^{N-k} \frac{\mathbf{P}\{\zeta_{N-k}^{(r)}=n-kr\}}{\mathbf{P}\{\zeta_N=n\}}. \tag{2}$$

PROOF. Let $A_{k,r}$ denote the event that exactly k of the random variables $\xi_1, \ldots, \xi_N$ assume the value r. From equation (1), we have

$$\mathbf{P}\{\mu_r(n, N)=k\}=\mathbf{P}\{A_{k,r} \mid \zeta_N=n\}=\frac{\mathbf{P}\{A_{k,r}, \zeta_N=n\}}{\mathbf{P}\{\zeta_N=n\}}.$$

Equation (2) results from obvious modifications of the numerator. Since the event $A_{k,r}$ may occur for C_N^k different choices of random variables assuming the value r, we can conclude that

$$\mathbf{P}\{A_{k,r}, \zeta_N = n\} = C_N^k\, p_r^k (1-p_r)^{N-k} \times$$
$$\times \mathbf{P}\{\zeta_N = n \mid \xi_1 \neq r, \ldots, \xi_{N-k} \neq r, \xi_{N-k+1} = r, \ldots, \xi_N = r\} =$$
$$= C_N^k p_r^k (1-p_r)^{N-k}\, \mathbf{P}\{\zeta_{N-k}^{(r)} = n - kr\}.$$

In (2), we now need only investigate $\mathbf{P}\{\zeta_{N-k}^{(r)} = n - kr\}$ since the investigation of the binomial probability and $\mathbf{P}\{\zeta_N = n\}$ is quite simple. If as $n, N \to \infty$ the values of α are in the interval $[\alpha_0, \alpha_1]$, where $0 < \alpha_0 \leqslant \alpha_1 < \infty$, the local theorem on convergence to a normal distribution can be applied to the sequence $\xi_1^{(r)}, \ldots, \xi_N^{(r)}$. In this case, equation (2) enables us to obtain for $\mu_r(n, N)$ the limit theorem on convergence to a normal distribution. This is apparently one of the simplest proofs of the asymptotic normality of $\mu_r(n, N)$ under the given conditions. If the range of variation of α covers the left-hand and the right-hand intermediate r-domains, the local theorem holds true for the sequence $\xi_1^{(r)}, \ldots, \xi_N^{(r)}$. This fact requires, however, a special proof because the general theory does not enable us to establish the validity of the local theorem for sequences even in this simple case.

Further, let us prove the local theorem on convergence to normal distributions of the sums $\zeta_N^{(r)}$, where $r \geqslant 2$, and, applying this theorem to (2), let us find the limit distributions for $\mu_r(n, N)$. Let us define

$$\alpha_r = \frac{\alpha - rp_r}{1 - p_r}, \quad \sigma_r^2 = \frac{\alpha}{(1-p_r)^2}\left(1 - p_r - \frac{(\alpha - r)^2}{\alpha} p_r\right).$$

It is easy to show that

$$\mathbf{M}\xi_1^{(r)} = \alpha_r, \qquad \mathbf{D}\xi_1^{(r)} = \sigma_r^2.$$

Theorem 1. *If* $m \longrightarrow \infty$ *and* $\alpha m \longrightarrow \infty$, *then, for* $r \geqslant 2$,

$$\mathbf{P}\{\zeta_m^{(r)} = l\} = \frac{1}{\sigma_r \sqrt{2\pi m}}\, e^{-\frac{(l - m\alpha_r)^2}{2m\sigma_r^2}} \left(1 + o(1)\right)$$

uniformly with respect to $\dfrac{l - m\alpha_r}{\sigma_r \sqrt{m}}$ *in any finite interval.*

PROOF. The characteristic function $\xi_1^{(r)}$ is equal to

$$f_r(t) = \frac{e^{\alpha(e^{it}-1)} - p_r e^{itr}}{1 - p_r}.$$

Let us define $f_r^*(t) = e^{-i\alpha_r t} f_r(t)$. The variance σ_r^2, as a function of α, is continuous and positive and $\sigma_r^2 = \alpha\left(1 + O\left(\left(\alpha + \frac{1}{\alpha}\right)p_r\right)\right)$ for $r \geqslant 2$ and any α. Hence, by the hypotheses of the theorem, $m\sigma_r^2 \to \infty$, and the integral theorem on convergence to a normal distribution holds true for the sequence $\xi_1^{(r)}, \ldots, \xi_m^{(r)}$, so that

$$\left(f_r^*\left(\frac{t}{\sigma_r \sqrt{m}}\right)\right)^m \to e^{-t^2/2} \tag{3}$$

uniformly with respect to t in any finite interval.

By direct calculation, we can easily verify that, for $\varphi_r(t) = \ln f_r^*(t)$,

$$\varphi_r(0) = 0, \quad \varphi_r'(0) = 0, \quad \varphi_r''(0) = -\sigma_r^2$$

and the variable $\left|\sigma_r^{-2} \varphi_r'''(t)\right|$ is bounded in a neighborhood of zero for any α. Hence, using a Taylor expansion, we find that

$$\varphi_r(t) = -\frac{\sigma_r^2 t^2}{2}(1 + O(t)).$$

From this it follows that

$$f_r^*(t) = \exp\left\{-\frac{\sigma_r^2 t^2}{2}(1 + O(t))\right\},$$

and therefore, there exist positive constants ε and c such that, for $0 \leqslant |t| \leqslant \varepsilon$,

$$|f_r^*(t)| \leqslant e^{-c\sigma_r^2 t^2}. \tag{4}$$

Let us estimate $|f_r(t)|$ in the domain $\varepsilon \leqslant |t| \leqslant \pi$ for other modes of variation of the parameter α. Note that

$$|f_r(t)| \leqslant \frac{e^{-\alpha(1-\cos t)} + p_r}{1-p_r} =$$

$$= e^{-\frac{\alpha(1-\cos t)}{3}} \frac{e^{-\frac{2}{3}\alpha(1-\cos t)} + \frac{\alpha^r}{r!} e^{-\frac{2}{3}\alpha\left(1+\frac{1}{2}\cos t\right)}}{1-p_r}.$$

For $\varepsilon \leqslant |t| \leqslant \pi$, the second factor approaches zero as $\alpha \to \infty$; hence, one can choose α_1 so that, for $\alpha > \alpha_1$ and $0 < \varepsilon \leqslant |t| \leqslant \pi$,

$$|f_r(t)| \leqslant e^{-\frac{\alpha(1-\cos t)}{3}} \leqslant e^{-\frac{\alpha\delta}{3}} = q^\alpha,$$

where $0 < \delta \leqslant 1-\cos\varepsilon$ and $q = e^{-\delta/3} < 1$.

As $\alpha \to 0$,

$$|f_r(t)| = e^{-\alpha(1-\cos t)}(1+O(\alpha^2)),$$

and one can choose α_0 so that, for $\alpha < \alpha_0$, and $\varepsilon \leqslant |t| \leqslant \pi$,

$$|f_r(t)| \leqslant e^{-\frac{\alpha(1-\cos t)}{3}} \leqslant q^\alpha.$$

For $\alpha_0 \leqslant \alpha \leqslant \alpha_1$, the estimate $|f_r(t)| \leqslant q^\alpha$ follows from the facts that, for any α, the maximum distribution step of $\xi_1^{(r)}$ is one, the characteristic function $f_r(t)$ depends continuously on α, and the set of the parameter values is compact. Therefore, for $0 < \varepsilon \leqslant |t| \leqslant \pi$, there exists a constant q in (0, 1) such that, no matter how α varies,

$$|f_r(t)| \leqslant q^\alpha. \tag{5}$$

Following the classical proof of the local theorem, we represent the probability $P_m(k) = \mathbf{P}\{\xi_1^{(r)} + \ldots + \xi_m^{(r)} = k\}$ as the integral

$$P_m(k) = \frac{1}{2\pi\sigma_r\sqrt{m}} \int_{-\pi\sigma_r\sqrt{m}}^{\pi\sigma_r\sqrt{m}} e^{-izt}\left(f_r^*\left(\frac{t}{\sigma_r\sqrt{m}}\right)\right)^m dt,$$

where $z = \dfrac{k - m\alpha_r}{\sigma_r\sqrt{m}}$. Since

$$\frac{1}{\sqrt{2\pi}} e^{-z^2/2} = \frac{1}{2\pi} \int_{-\infty}^{\infty} e^{-izt-\frac{t^2}{2}}\, dt,$$

the difference $R_m = 2\pi\left[\sigma_r\sqrt{m}\,P_m(k) - \dfrac{1}{\sqrt{2\pi}} e^{-z^2/2}\right]$ is representable as the sum of four integrals:

$$R_m = I_1 + I_2 + I_3 + I_4,$$

where

$$I_1 = \int_{-A}^{A} e^{-izt}\left[\left(f_r^*\left(\frac{t}{\sigma_r\sqrt{m}}\right)\right)^m - e^{-t^2/2}\right] dt,$$

$$I_2 = \int_{A<|t|\leqslant\varepsilon\sigma_r\sqrt{m}} e^{-izt}\left(f_r^*\left(\frac{t}{\sigma_r\sqrt{m}}\right)\right)^m dt,$$

$$I_3 = \int_{\varepsilon\sigma_r\sqrt{m}<|t|\leqslant\pi\sigma_r\sqrt{m}} e^{-izt}\left(f_r^*\left(\frac{t}{\sigma_r\sqrt{m}}\right)\right)^m dt,$$

$$I_4 = -\int_{A<|t|} e^{-izt-\frac{t^2}{2}}\, dt.$$

For $|I_4|$ we have the estimate

$$|I_4| \leqslant \int_{A<|t|} e^{-t^2/2} dt.$$

Using the inequality given in (4) for estimating the integral I_2, we find that

$$|I_2| \leqslant \int\limits_{A\leqslant|t|\leqslant\varepsilon\sigma_r\sqrt{m}} \left| f_r^m\left(\frac{t}{\sigma_r\sqrt{m}}\right)\right| dt \leqslant \int\limits_{A\leqslant|t|} e^{-ct^2}\,dt.$$

Therefore, the variables $|I_2|$ and $|I_4|$ can be made as small as desired by choosing A sufficiently large. For fixed A, the integral $I_1 \to 0$ as $m \to \infty$ by virtue of (3). Finally, as $m \to \infty$,

$$|I_3| \leqslant \sigma_r\sqrt{m} \int\limits_{\varepsilon\leqslant|t|\leqslant\pi} |f_r^m(t)|\,dt \leqslant \sigma_r\sqrt{m}\, q^{\alpha m} \to 0\,,$$

since, as noted above, $\sigma_r^2 = O(\alpha)$ for arbitrary variation of α and as $\alpha m \to \infty$. The theorem is proved.

We use Lemma 1 and Theorem 1 for investigating limit distributions of $\mu_r(n, N)$, where $r \geqslant 2$.

Theorem 2. *If $r \geqslant 2$ is fixed and $n, N \to \infty$ in such a way that $Np_r \to \infty$, then*

$$\mathbf{P}\{\mu_r(n, N) = k\} = \frac{1}{\sigma_{rr}\sqrt{2\pi N}}\, e^{-\frac{(k-Np_r)^2}{2N\sigma_{rr}^2}}(1 + o(1))$$

uniformly with respect to $\dfrac{k-Np_r}{\sigma_{rr}\sqrt{N}}$ *in any finite interval* (σ_{rr} *is given by* (1.9)).

PROOF. Using the normal approximation for a binomial distribution as $Np_r(1-p_r) \to \infty$, we find that

$$C_N^k p_r^k (1-p_r)^{N-k} = \frac{1}{\sqrt{2\pi Np_r(1-p_r)}}\, e^{-\frac{(k-Np_r)^2}{2Np_r(1-p_r)}}(1+o(1)) \tag{6}$$

uniformly with respect to $\dfrac{k-Np_r}{\sqrt{Np_r(1-p_r)}}$ in any finite interval and hence uniformly with respect to $\dfrac{k-Np_r}{\sigma_{rr}\sqrt{N}}$ since $\sigma_{rr}^2 = p_r\left(1-p_r-\dfrac{(\alpha-r)^2}{\alpha}p_r\right) \leqslant p_r(1-p_r)$.

The sum ζ_N is Poisson-distributed with parameter n; hence, as $n \to \infty$,

$$\mathbf{P}\{\zeta_N = n\} = \frac{n^n e^{-n}}{n!} = \frac{1}{\sqrt{2\pi n}}(1 + o(1)). \tag{7}$$

To estimate the probability $\mathbf{P}\{\zeta_{N-k}^{(r)} = n - kr\}$, we use Theorem 1. We define $u = \frac{k - Np_r}{\sigma_{rr}\sqrt{N}}$. Note that, for bounded $|u|$, we have

$$N - k = N(1 - p_r)\left(1 - \frac{u\sigma_{rr}}{(1 - p_r)\sqrt{N}}\right) = N(1 - p_r)(1 + o(1)).$$

Setting $m = N - k$ and $l = n - rk$ in Theorem 1 and substituting the expressions

$$\alpha_r = \frac{\alpha - rp_r}{1 - p_r}, \qquad \sigma_r^2 = \frac{\alpha\sigma_{rr}^2}{p_r(1 - p_r)^2},$$

we find that

$$\frac{(l - m\alpha_r)^2}{2m\sigma_r^2} = \frac{(\alpha - r)^2 p_r u^2}{2\alpha(1 - p_r)}(1 + o(1))$$

uniformly with respect to u in any finite interval.

Since $\frac{(\alpha - r)^2 p_r}{\alpha(1 - p_r)}$ is bounded, it follows from Theorem 1 that

$$\mathbf{P}\{\zeta_{N-k}^{(r)} = n - kr\} = \frac{1}{\sigma_r\sqrt{2\pi N(1 - p_r)}} e^{-\frac{(\alpha - r)^2 p_r u^2}{2\alpha(1 - p_r)}}(1 + o(1)) \tag{8}$$

uniformly with respect to u in any finite interval. To complete the proof, it only remains to substitute the estimates given by (6), (7), and (8) into (2).

Theorem 3. *If n, $N \to \infty$ in such a way that $\alpha \to 0$ or $\alpha \to \infty$, then for any fixed $r \geqslant 2$,*

$$\mathbf{P}\{\mu_r(n, N) = k\} = \frac{1}{k!}(Np_r)^k e^{-Np_r}(1 + o(1))$$

uniformly with respect to $(k - Np_r)/\sqrt{Np_r}$ in any finite interval.

PROOF. Since $p_r \to 0$ if the conditions of the theorem are satisfied, it follows that

$$C_N^k p_r^k (1 - p_r)^{N-k} = \frac{1}{k!}(Np_r)^k e^{-Np_r}(1 + o(1)) \qquad (9)$$

uniformly with respect to $(k - Np_r)/\sqrt{Np_r}$ in any finite interval.

We shall show that if the conditions of the theorem are satisfied, the ratio of the probabilities $\mathbf{P}\{\zeta_{N-k}^{(r)} = n - kr\}/\mathbf{P}\{\zeta_N = n\}$ tends to 1. The estimate (7) holds for $\mathbf{P}\{\zeta_N = n\}$. If $(k - Np_r)/\sqrt{Np_r}$ is bounded, then $N - k = N(1 + o(1))$ and, for $m = N - k$ and $l = n - kr$,

$$\frac{(l - m\alpha_r)^2}{2m\sigma_r^2} \to 0.$$

Hence, from Theorem 1 we infer that

$$\mathbf{P}\{\zeta_{N-k}^{(r)} = n - kr\} = \frac{1}{\sqrt{2\pi\alpha N}}(1 + o(1)). \qquad (10)$$

Substituting the estimates given by (9), (7), and (10) into (2), we arrive at the assertion of the theorem.

From the local Theorem 2, we get in the usual way the integral theorem on convergence of the distribution $\dfrac{\mu_r(n, N) - Np_r}{\sigma_{rr}\sqrt{N}}$ to a normal distribution with parameters (0, 1). It follows from Theorem 1.1 that for any $r \geqslant 2$, as n, N, $Np_r \to \infty$

$$\mathbf{M}\mu_r(n, N) = Np_r + O(1), \quad \mathbf{D}\mu_r(n, N) = N\sigma_{rr}^2(1 + (o(1)).$$

Therefore, this assertion can be formulated as the following:

Theorem 4. *If* $r \geqslant 2$ *is fixed, then*

$$\mathbf{P}\left\{\frac{\mu_r(n, N) - \mathbf{M}\mu_r(n, N)}{\sqrt{\mathbf{D}\mu_r(n, N)}} \leqslant x\right\} \to \frac{1}{\sqrt{2\pi}} \int_{-\infty}^{x} e^{-u^2/2} du.$$

with respect to x *in* $(-\infty, \infty)$ *as* $n, N \to \infty$ *in such a way that* $Np_r \to \infty$.

Since $Np_r \to \lambda$ in the left-hand and right-hand r-domains, from Theorem 3 we get

Theorem 5. *In the left-hand and right-hand* r*-domains, for any fixed* $k=0, 1, 2, \ldots$,

$$\mathbf{P}\{\mu_r(n, N) = k\} \to \frac{\lambda^k e^{-\lambda}}{k!}$$

as $n, N \to \infty$.

In Theorems 2 and 3, we examined the asymptotic behavior of the random variables $\mu_r(n, N)$ for arbitrary fixed $r \geqslant 2$. If the parameter $r \to \infty$ as $n, N \to \infty$, the distribution of $\mu_r(n, N)$ approaches a Poisson distribution with parameter Np_r:

Theorem 6.

$$\mathbf{P}\{\mu_r(n, N) = k\} = \frac{1}{k!}(Np_r)^k e^{-Np_r}(1 + o(1))$$

uniformly with respect to $(k - Np_r)/\sqrt{Np_r}$ *in any finite interval as* $n, N, r \to \infty$.

PROOF. Since the estimates given by (7) and (9) hold, to prove the theorem it suffices to verify that the estimate given by (10) holds as $r \to \infty$. If $(k - Np_r)/\sqrt{Np_r}$ is bounded, under the conditions of the theorem $N - k = N(1 + o(1))$, and, for $m = N - k$ and $l = n - kr$, we have

$$\frac{(l - m\alpha_r)^2}{2m\sigma_r^2} \to 0, \tag{11}$$

since $\frac{(\alpha - r)^2}{\alpha} p_r \to 0$ for arbitrary variation of α as $r \to \infty$. In fact, the function $f(\alpha) = \frac{(\alpha - r)^2}{\alpha} p_r$ attains a maximum with respect to α at $\alpha^* = r + \sqrt{r}$, but $f(\alpha^*) \to 0$ as $\alpha^* \to \infty$. By virtue of Theorem 1, the estimate given by (10) follows from (11), thus proving the theorem.

§4. Limit Distributions of μ_1

This section deals with the limit distributions of the number $\mu_1(n, N)$ of cells containing exactly one particle. The reason we single out the results dealing with $\mu_1(n, N)$ is that the behavior of this random variable as $\alpha \to 0$ has some noteworthy features.

To study the limit distributions of $\mu_1(n, N)$, we use Lemma 3.1. Equation (3.2) in this case becomes

$$\mathbf{P}\{\mu_1(n, N) = k\} = C_N^k p_1^k (1 - p_1)^{N-k} \frac{\mathbf{P}\{\zeta_{N-k}^{(1)} = n - k\}}{\mathbf{P}\{\zeta_N = n\}}, \tag{1}$$

where $\zeta_N = \xi_1 + \ldots + \xi_N$, $\zeta_N^{(1)} = \xi_1^{(1)} + \ldots + \xi_N^{(1)}$, the random variables $\xi_1, \ldots, \xi_N$ and $\xi_1^{(1)}, \ldots, \xi_N^{(1)}$ are independent and identically distributed, and

$$\mathbf{P}\{\xi_1 = k\} = p_k = \frac{\alpha^k e^{-\alpha}}{k!}, \qquad k = 0, 1, \ldots,$$

$$\mathbf{P}\{\xi_1^{(1)} = k\} = \mathbf{P}\{\xi_1 = k \mid \xi_1 \neq 1\}.$$

As in investigating $\mu_r(n, N)$, where $r \geqslant 2$, we show here that the local limit theorem holds for the sum $\zeta_{N-k}^{(1)}$. Denote

$$\alpha_1 = \frac{\alpha - p_1}{1 - p_1}, \qquad \sigma_1^2 = \frac{\alpha}{(1 - p_1)^2}\left(1 - p_1 - \frac{(\alpha - 1)^2}{\alpha} p_1\right).$$

It is easy to see that $\alpha_1 = \mathbf{M}\xi_1^{(1)}$, $\sigma_1^2 = \mathbf{D}\xi_1^{(1)}$.

Theorem 1. *If $m \to \infty$ and $\alpha^3 m \to \infty$, then*

$$\mathbf{P}\{\zeta_m^{(1)} = l\} = \frac{1}{\sigma_1 \sqrt{2\pi m}} e^{-\frac{(l-m\alpha_1)^2}{2m\sigma_1^2}} (1 + o(1))$$

uniformly with respect to $\frac{l - m\alpha_1}{\sigma_1 \sqrt{m}}$ *in any finite interval.*

PROOF. The characteristic function $\xi_i^{(1)}$ is equal to

$$f_1(t) = \frac{e^{\alpha(e^{it}-1)} - p_1 e^{it}}{1 - p_1}.$$

Denote $f_1^*(t) = e^{-it\alpha_1} f_1(t)$. The behavior of the sum $\zeta_m^{(1)}$ has the peculiarity that in estimating

$$R_m = 2\pi \left[\sigma_1 \sqrt{m}\, \mathbf{P}\{\zeta_m^{(1)} = l\} - \frac{1}{\sqrt{2\pi}} e^{-z^2/2}\right],$$

where $z = \frac{l - m\alpha_1}{\sigma_1 \sqrt{m}}$, in the case in which $\alpha \to 0$, the effect of the local maximum $|f_1(t)|$ at the point π must be considered. Hence, in contrast to the case $r \geqslant 2$, we isolate a domain containing neighborhoods of the points π and $-\pi$. Let us also represent R_m as the sum of five integrals:

$$R_m = I_1 + I_2 + I_3 + I_4 + I_5,$$

where

$$I_1 = \int_{-A}^{A} e^{-izt} \left[\left(f_1^*\left(\frac{t}{\sigma_1 \sqrt{m}}\right)\right)^m - e^{-t^2/2}\right] dt,$$

$$I_2 = \int_{A \leqslant |t| \leqslant \varepsilon\sigma_1\sqrt{m}} e^{-izt} \left(f_1^*\left(\frac{t}{\sigma_1 \sqrt{m}}\right)\right)^m dt,$$

$$I_3 = \int_{\varepsilon\sigma_1\sqrt{m} \leqslant |t| \leqslant (\pi-\varepsilon)\sigma_1\sqrt{m}} e^{-izt} \left(f_1^*\left(\frac{t}{\sigma_1 \sqrt{m}}\right)\right)^m dt,$$

$$I_4 = -\int_{A \leqslant |t|} e^{-izt - t^2/2}\, dt,$$

$$I_5 = \int_{(\pi-\varepsilon)\sigma_1\sqrt{m} \leqslant |t| \leqslant \pi\sigma_1\sqrt{m}} e^{-izt} \left(f_1^*\left(\frac{t}{\sigma_1 \sqrt{m}}\right)\right)^m dt.$$

The variance σ_1^2 as a function of α is continuous and positive. Also, no matter how α varies,

$$\sigma_1^2 = \alpha(1 - e^{-2\alpha})(1 + O((1 + \alpha) p_1)). \tag{2}$$

Hence, under the conditions of the theorem, $m\sigma_1^2 \to \infty$ and the integral theorem on convergence to the normal law is applicable for the sequence $\xi_1^{(1)}, \ldots, \xi_m^{(1)}$, so that

$$\left(f_1^*\left(\frac{t}{\sigma_1 \sqrt{m}}\right)\right)^m \to e^{-t^2/2} \tag{3}$$

uniformly with respect to t in any finite interval.

Since

$$|f_1(t)| \leqslant \frac{e^{-\alpha(1-\cos t)} + p_1}{1 - p_1} =$$

$$= e^{-\frac{\alpha(1-\cos t)}{3}} \frac{e^{-\frac{2}{3}\alpha(1-\cos t)} + \alpha e^{-\frac{2}{3}\alpha\left(1+\frac{1}{2}\cos t\right)}}{1 - p_1}$$

and since, for $\varepsilon \leqslant |t| \leqslant \pi$, the second factor approaches zero as $\alpha \to \infty$, one can choose α_1 such that, for $\alpha > \alpha_1$ and $\varepsilon \leqslant |t| \leqslant \pi$,

$$|f_1(t)| \leqslant e^{-\frac{\alpha(1-\cos t)}{3}} \leqslant q^{\frac{\alpha^2}{1+\alpha}}, \quad q < 1. \tag{4}$$

In the limit as $\alpha \to 0$,

$$|f_1(t)| = \frac{\left|e^{\alpha e^{it}} - \alpha e^{it}\right|}{e^{\alpha} - \alpha} = 1 - \frac{\alpha^2}{2}(1 - \cos 2t) + O(\alpha^3).$$

Hence, there exist α_0 and q in (0, 1) such that, for $\alpha < \alpha_0$ and $\varepsilon \leqslant |t| \leqslant \pi - \varepsilon$,

$$|f_1(t)| \leqslant q^{\alpha^2/(1+\alpha)}.$$

Finally, for $\alpha_0 \leqslant \alpha \leqslant \alpha_1$ the estimate of the characteristic function

$$|f_1(t)| \leqslant q^{\alpha^2/(1+\alpha)}, \qquad 0 < q < 1,$$

follows from the facts that the maximum step of the distributions is equal to one and $f_1(t)$ depends continuously on the parameter α. Therefore, for $\varepsilon \leqslant |t| \leqslant \pi - \varepsilon$, there is a constant q in (0, 1) such that

$$|f_1(t)| \leqslant q^{\alpha^2/(1+\alpha)}. \tag{5}$$

As in the proof of Theorem 3.1, we find here that the integral I_4 can be made as small as desired by choosing A sufficiently large. By virtue of (3) and (5), the integrals I_1 and I_3 approach zero as $m \to \infty$.

It remains to estimate the integrals I_2 and I_5. As in the case $r \geqslant 2$, it is easy to verify by direct calculation that, for the function $\varphi_1(t) = \ln f_1^*(r)$,

$$\varphi_1(0) = 0, \qquad \varphi_1'(0) = 0, \qquad \varphi_1''(0) = -\sigma_1^2,$$

and the variable $|\sigma_1^{-2} \varphi_1'''(t)|$ is bounded in a neighborhood of zero and in a neighborhood of π. Furthermore, for sufficiently small α,

$$\varphi_1(\pi) < -2c_1\alpha^3, \qquad \varphi_1'(\pi) = ic(\alpha), \qquad \varphi_1''(\pi) < -2c_2\sigma_1^2,$$

where c_1 and c_2 are positive constants and $c(\alpha)$ is a real function.

From this it follows that there exist positive c and ε such that, for $0 \leqslant |t| \leqslant \varepsilon$,

$$|f_1(t)| \leqslant e^{-c\sigma_1^2 t^2} \tag{6}$$

and, for $|\pi - t| \leqslant \varepsilon$,

$$|f_1(t)| \leqslant e^{-c_1\alpha^3 - \frac{1}{2}c_2\sigma_1^2(\pi - t)^2} \tag{7}$$

Using (6) to estimate the integral I_2, we find that

$$|I_2| \leqslant \int\limits_{A \leqslant |t|} e^{-ct^2} dt,$$

and this integral can be made as small as desired by choosing A sufficiently large.

If $\alpha \geqslant \alpha_0 > 0$, we can use (4) to estimate I_5. Then we find that

$$|I_5| \leqslant 2\sigma_1 \sqrt{m} \int\limits_{\pi-\varepsilon}^{\pi} |f_1^m(t)| \, dt \leqslant 2\varepsilon\sigma_1 \sqrt{m}\, q^{m\alpha^2/(1+\alpha)}$$

and, by virtue of (2), the right side approaches zero for $\alpha \geqslant \alpha_0 > 0$ as $m \to \infty$. If $\alpha \to 0$, we use (7) to estimate the integral I_5. Because of the periodicity of the characteristic function $f_1(t)$,

$$|I_5| \leqslant \int\limits_{(\pi-\varepsilon)\sigma_1\sqrt{m} \leqslant |x| \leqslant \pi\sigma_1\sqrt{m}} \left| f_1^m\left(\frac{x}{\sigma_1\sqrt{N}}\right)\right| dx =$$

$$= \int\limits_{(\pi-\varepsilon)\sigma_1\sqrt{m}}^{(\pi+\varepsilon)\sigma_1\sqrt{m}} \left| f_1^m\left(\frac{x}{\sigma_1\sqrt{m}}\right)\right| dx = \sigma_1 \sqrt{m} \int\limits_{-\varepsilon}^{\varepsilon} |f_1^m(\pi+t)| \, dt.$$

Using (7) to estimate the last integral, we find that

$$|I_5| \leqslant \sigma_1 \sqrt{m} e^{-\alpha^3 m} \int\limits_{-\varepsilon}^{\varepsilon} e^{-\frac{1}{2} c_2 \sigma_1^2 t^2 m} dt \leqslant e^{-\alpha^3 m} \int\limits_{-\infty}^{\infty} e^{-\frac{1}{2} c_2 u^2} du.$$

From this it follows that $I_5 \to 0$ as $\alpha^3 m \to \infty$, thus proving Theorem 1.

In the notation introduced in Section 1, we write $\sigma_{11}^2 = p_1\left(1 - p_1 - \frac{(\alpha-1)^2}{\alpha} p_1\right)$.

Applying Theorem 1 and equation (1), we can deduce the assertions given below in the same way as in investigating the limit distributions of $\mu_r(n, N)$, where $r \geqslant 2$.

Theorem 2. *If $n, N \to \infty$ in such a way that $\alpha^3 N \to \infty$ and $Np_1 \to \infty$, then*

$$\mathbf{P}\{\mu_1(n, N) = k\} = \frac{1}{\sigma_{11}\sqrt{2\pi N}} e^{-\frac{(k-Np_1)^2}{2N\sigma_{11}^2}} (1 + o(1))$$

uniformly with respect to $\dfrac{k - Np_1}{\sigma_{11}\sqrt{N}}$ in any finite interval.

Theorem 3. *If $n, N, \alpha \to \infty$, then*

$$\mathbf{P}\{\mu_1(n, N) = k\} = \frac{1}{k!}(Np_1)^k e^{-Np_1}(1 + o(1))$$

uniformly with respect to $(k-Np_1)/\sqrt{Np_1}$ in any finite interval.

The proofs of these theorems follow step by step the proofs of Theorems 3.2 and 3.3 if we set $r=1$ in them.

Let us consider now the behavior of $\mu_1(n, N)$ as $\alpha \to 0$. If $\alpha \to 0$ and $\alpha^3 N \to \infty$, then, by Theorem 2, the distribution of $\mu_1(n, N)$ approximates a normal distribution locally. For $\mu_r(n, N)$, where $r \geqslant 2$, the parameters of the limiting normal distribution in the left-hand intermediate r-domain are asymptotically identical and equal to Np_r, so that in the domain where $\alpha \to 0$ and $Np_r \to \infty$, the distribution of $\mu_r(n, N)$ also approximates a Poisson distribution. This is not true for $\mu_1(n, N)$. Using Theorem 2, we can obtain the following result. Let

$$\lambda = \frac{1}{2}\alpha(1 - e^{-\alpha})N.$$

Theorem 4. *If $n, N \to \infty$ in such a way that $\alpha \to 0$ and $\alpha^3 N \to \infty$, then*

$$\mathbf{P}\{\mu_1 = n - 2k\} = \frac{1}{2}\frac{\lambda^k e^{-\lambda}}{k!}(1 + o(1)), \tag{8}$$

$$\mathbf{P}\{\mu_1 = n - 2k - 1\} = \frac{1}{2}\frac{\lambda^k e^{-\lambda}}{k!}(1 + o(1)) \tag{9}$$

uniformly with respect to $(k-\lambda)/\sqrt{\lambda}$ *in any finite interval.*

PROOF. Since a Poisson distribution with parameter λ approximates a normal distribution as $\lambda \to \infty$ and since $\lambda = \frac{1}{2}\alpha(1 - e^{-\alpha})N \to \infty$, it follows that

$$\frac{1}{2}\frac{\lambda^k e^{-\lambda}}{k!} = \frac{1}{2\sqrt{2\pi\lambda}} e^{-\frac{(k-\lambda)^2}{2\lambda}}(1 + o(1)) \tag{10}$$

uniformly with respect to $(k-\lambda)/\sqrt{\lambda}$. On the other hand, by Theorem 2, we infer that

$$\mathbf{P}\{\mu_1 = n - 2k\} = \frac{1}{\sigma_{11}\sqrt{2\pi N}} e^{-\frac{(n-2k-Np_1)^2}{2N\sigma_{11}^2}}(1 + o(1)) \tag{11}$$

uniformly with respect to $\dfrac{n - 2k - Np_1}{\sigma_{11}\sqrt{N}}$ on any finite interval. Noting that

$$N\sigma_{11}^2 = 4\lambda(1 + O(\alpha)),$$

we obtain from (11) the result that

$$\mathbf{P}\{\mu_1 = n - 2k\} = \frac{1}{2\sqrt{2\pi\lambda}} e^{-\frac{(k-\lambda)^2}{2\lambda}}(1 + o(1)) \tag{12}$$

uniformly with respect to $(k-\lambda)/\sqrt{\lambda}$ in any finite interval.

Equation (8) follows from (10) and (12). Equation (9) is proved in a similar way.

If $\alpha \to 0$ and $\alpha^3 N$ is bounded, there is a simple connection between the asymptotic behavior of the random variable $\mu_1(n, N)$ and the behavior of $\mu_2(n, N)$. It follows from (1.6) that

$$\mathbf{M}\mu_r(n, N) \leqslant N \frac{\alpha^r}{r!}.$$

Using this estimate, we have

$$\mathbf{M} \sum_{r \geqslant 3} r\mu_r(n, N) \leqslant \alpha^3 N \sum_{r \geqslant 3} \frac{\alpha^{r-3}}{(r-1)!} \leqslant \alpha^3 N e^{\alpha}. \tag{13}$$

From the equation $\mu_1 + 2\mu_2 + \ldots + n\mu_n = n$ it follows that

$$\frac{n - \mu_1(n, N)}{2} = \mu_2(n, N) + \frac{1}{2} \sum_{r \geqslant 3} r\mu_r(n, N). \tag{14}$$

with the aid of (13) and (14), we easily arrive at the following assertions:

Theorem 5. *If n, $N \to \infty$ in such a way that $\alpha^2 N \to \infty$, then*

$$\mathbf{P}\left\{\frac{\mu_1(n, N) - Np_1}{\sigma_{11}\sqrt{N}} \leqslant x\right\} \to \frac{1}{2\pi} \int_{-\infty}^{x} e^{-u^2/2} du$$

uniformly with respect to x.

PROOF. If $\alpha^3 N \to \infty$, the assertion of the theorem follows from the local Theorem 2 in the usual way. Hence, we consider the case where $\alpha^2 N \to \infty$ and $\alpha^3 N$ is bounded. Using the relation given in (14), we find that

$$\frac{Np_2 - \frac{1}{2}(n - \mu_1)}{\sqrt{Np_2}} = \frac{Np_2 - \mu_2}{\sqrt{Np_2}} - \frac{1}{\sqrt{Np_2}} \sum_{r \geqslant 3} r\mu_r. \tag{15}$$

Since, for bounded $\alpha^3 N$, the random variable $(Np_2)^{-1/2} \sum_{r \geqslant 3} r\mu_r$ tends to zero in probability and $(Np_2)^{-1/2}(Np_2 - \mu_2)$ is asymptotically normal, it follows from (15) that

$$\mathbf{P}\left\{\frac{Np_2 - \frac{1}{2}(n-\mu_1)}{\sqrt{Np_2}} \leqslant x\right\} \to \frac{1}{\sqrt{2\pi}} \int_{-\infty}^{x} e^{-u^2/2} du. \tag{16}$$

Noting that

$$Np_1 = \alpha N - \alpha^2 N + O(\alpha^3 N), \qquad \sigma_{11}^2 = 2\alpha^2(1 + O(\alpha)),$$

if $\alpha \to 0$ as $n,\ N \to \infty$ and $\alpha \to 0$, we see that, if $\alpha^2 N \to \infty$ but $\alpha^3 N$ is bounded, the random variable $(Np^2 - {}^1/_2(n - \mu_1))/\sqrt{Np_2}$ tends to $\dfrac{\mu_1 - Np_1}{\sigma_{11}\sqrt{N}}$ in probability. The assertion of the theorem now follows from (16) in a domain where $\alpha^2 N \to \infty$ and $\alpha^3 N$ is bounded.

Theorem 6. *If* $n, N \to \infty$ *in such a way that* $\alpha \to 0$ *and* $Np_2 \to \lambda$, *where* λ *is a positive constant, then, for any fixed* $k = 0, 1, 2, \ldots,$

$$\mathbf{P}\left\{\frac{n - \mu_1(n, N)}{2} = k\right\} \to \frac{\lambda^k e^{-\lambda}}{k!}.$$

PROOF. By Theorem 3.5, the random variable $\mu_2(n, N)$ has in the limit a Poisson distribution with parameter λ if the conditions of Theorem 6 are satisfied. By virtue of (13), the random variable $\sum_{r \geqslant 3} r\mu_r$, converges to zero in probability as $\alpha^3 N \to 0$. Hence it follows from (14) that $(n - \mu_1)/2$ and μ_2 have identical limit distributions. Therefore, as $Np_2 \to \lambda$, the distribution of $n - \mu_1$ is concentrated asymptotically on the lattice of nonnegative even numbers. It follows from (13) and (14) that this property holds as $\alpha^3 N \to 0$. This fact explains why the local Theorem 2 on convergence to a normal distribution on the lattice of all integers does not hold true in this domain whereas the integral theorem holds true in a wider domain as $\alpha^2 N \to \infty$. As $\alpha^3 N \to \infty$, the distribution of $\mu_1(n, N)$ is, by virtue of Theorem 2, concentrated on all integral points. If the values of the parameter $\alpha^3 N$ are bounded, the distribution of the random variable $n - \mu_1$ jumps from the lattice of all nonnegative integers to a lattice of nonnegative even numbers. We shall not discuss this jump phenomenon here.

§5. The Rate of Convergence to Limit Distributions

In Section 3, we showed that the distributions of the random variables $\mu_r(n, N)$, where $r \geqslant 2$, approximate a normal and a Poisson distribution in the left- and right-hand intermediate r-domains. It makes sense to estimate the proximity of the distribution of $\mu_r(n, N)$ to the limit distributions in these domains and to indicate which one of the limit distributions provides the closest approximation for a given mode of variation of the parameter α.

In the present section, we shall use refinements of the local limit Theorems 3.1 and 3.2 to find the distances with respect to variation between the distributions of $\mu_r(n, N)$ and the limit distributions. This enables us to indicate the limit distribution closest to the distribution of $\mu_r(n, N)$ in the sense of this distance and to calculate an upper bound, in terms of α, of the minimum distance.

The behavior of the distributions of $\mu_r(n, N)$ with $r \geqslant 2$ is similar to that of a binomial distribution: as $n \to \infty$, the binomial distribution can, depending on the way the parameter p varies, approximate a normal or a Poisson distribution. We discuss first certain results dealing with the approximation of the binomial distribution since the problem of the rate of convergence of distributions of $\mu_r(n, N)$ to limit distributions may be investigated in much the same manner as in the classical case.

The validity of the theorems which follow can be established by direct calculations with the aid of Stirling's formula.

Theorem 1. *If* $\dfrac{1+u^6}{npq} \to 0$ *as* $n \to \infty$, *where* $q = 1 - p$ *and* $u = (k - np)/\sqrt{npq}$, *then*

$$C_n^k p^k q^{n-k} = \frac{1}{\sqrt{2\pi npq}} e^{-u^2/2}\left(1 + \frac{q-p}{6\sqrt{npq}}(3u - u^3) + O\left(\frac{1+u^6}{npq}\right)\right).$$

Theorem 2. *If* $np \to \infty$ *and* $(1+u^2)p \to 0$ *as* $n \to \infty$, *where* $u = (k - np)/\sqrt{np}$, *u* $q=1-p$, *then*

$$C_n^k\, p^k q^{n-k} = \frac{1}{k!}(np)^k e^{-np}\left(1 + \frac{1}{2}\,p\,(1-u^2) + \right.$$
$$\left. + O\left(p^2(1+u^4) + \frac{(1+|u|)\,p}{\sqrt{np}}\right)\right)$$

Theorem 3. *If* $(1+u^6)/\lambda \to 0$, *where* $u = (k-\lambda)/\sqrt{\lambda}$, *then*

$$\frac{\lambda^k e^{-\lambda}}{k!} = \frac{1}{\sqrt{2\pi\lambda}}\, e^{-u^2/2}\left(1 + \frac{u^3 - 3u}{6\sqrt{\lambda}} + O\left(\frac{1+u^6}{\lambda}\right)\right).$$

Here and below, an expression of the form $f(u_1, \ldots, u_n) = O(\varphi(u_1, \ldots, u_n))$ means that, with the parameter behavior indicated, $|f| \leqslant C|\varphi|$, where C is some absolute constant.

Let us define

$$\Pi(k, \lambda) = \frac{\lambda^k e^{-\lambda}}{k!}, \qquad k = 0, 1, \ldots,$$
$$N(x, a, \sigma^2) = \frac{1}{\sigma\sqrt{2\pi}}\, e^{-(x-a)^2/2\sigma^2}. \tag{1}$$

Let us define distances between binomial distributions and limit distributions as follows:

$$\rho_1(n, p) = \sum_{k=0}^{n} \left|C_n^k\, p^k q^{n-k} - \Pi(k, np)\right|,$$
$$\rho_2(n, p) = \sum_{k=0}^{n} \left|C_n^k\, p^k q^{n-k} - N(k, np, npq)\right|,$$
$$\rho_3(n, p) = \rho_1(n, q).$$

To evaluate these distances asymptotically, we shall need three lemmas, which we shall use in replacing summation with integration.

Lemma 1. *Suppose that a bounded continuous function* $\varphi(x)$ *is* *integrable on* $(-\infty, \infty)$ *and has a finite number* s *of maxima and minima. Let us define* $x_k = x_0 + k\Delta$, *where* k *is an integer and* $\Delta > 0$. *Then, for any* c *and* d *such that* $c < d$,

$$\left|\int_c^d \varphi(x)\,dx - \sum_{x_k \in [c,d)} \varphi(x_k)\,\Delta\right| \leqslant 2(s+1) \sup_{c \leqslant x < d} |\varphi(x)|\,\Delta.$$

PROOF. On every interval $[l, h) \subset [c, d)$ on which $\varphi(x)$ is monotone, the difference

$$\left|\int_l^h \varphi(x)\,dx - \sum_{x_k \in [l,h)} \varphi(x_k)\,\Delta\right|$$

does not exceed $2 \sup_{l \leqslant x < h} |\varphi(x)|\,\Delta$, and therefore, the difference

$$\left|\int_c^d \varphi(x)\,dx - \sum_{x_k \in [c,d)} \varphi(x_k)\,\Delta\right|$$

does not exceed $2(s+1) \sup_{c \leqslant x < d} |\varphi(x)|\,\Delta$.

Lemma 2. *Let* $\varphi(x)$ *denote an even function that is integrable on* $(-\infty, \infty)$. *Suppose that* $\varphi(x)$ *is nonincreasing on* $[0, \infty)$ *and that its first and second derivatives are uniformly bounded. Let* T *be a positive integer. Define* $x_k = x_0 + k\Delta$, *where* k *assumes integer values and* $\Delta > 0$. *Then, for* $c < 0 < d$,

$$\sum_{x_k \in [c,d)} \varphi(x_k)\,\Delta = \int_c^d \varphi(x)\,dx + O(\Delta^3 T + \varphi(T\Delta)\,\Delta)$$

uniformly with respect to T *and* Δ *such that* $T\Delta \leqslant \min(-c, d)$.

PROOF. Consider first the case where $x_0 = -m\Delta$ for some integer m. Applying Lemma 1, we find that

$$\sum_{x_k \in [c,d)} \varphi(x_k)\Delta =$$
$$= \sum_{x_k \in [-T\Delta, T\Delta)} \varphi(x_k)\Delta + \sum_{x_k \in [c, -T\Delta)} \varphi(x_k)\Delta +$$
$$+ \sum_{x_k \in [T\Delta, d)} \varphi(x_k)\Delta = \sum_{x_k \in [-T\Delta, T\Delta)} \varphi(x_k)\Delta +$$
$$+ \int_c^{-T\Delta} \varphi(x)\,dx + \int_{T\Delta}^{d} \varphi(x)\,dx + O(\varphi(T\Delta)\Delta). \quad (2)$$

Treating the integrals

$$a_t = \int_{x_{m-t-1}}^{x_{m-t}} \varphi(x)\,dx + \int_{x_{m+t}}^{x_{m+t+1}} \varphi(x)\,dx$$

as functions of the upper limits and expanding them in Taylor series up to terms of order Δ^3, we find that, for any integer t,

$$a_t = (\varphi(x_{m-t-1}) + \varphi(x_{m+t}))\Delta +$$
$$+ \frac{\Delta^2}{2}(\varphi'(x_{m-t-1}) + \varphi'(x_{m+t})) + O(\Delta^3).$$

Note that $x_{m-t-1} = -\Delta(t+1)$, $x_{m+t} = t\Delta$, and $\varphi'(-\Delta(t+1)) = -\varphi'(t\Delta+\Delta) = -\varphi'(t\Delta) + O(\Delta)$. Hence, $a_t = (\varphi(x_{m-t-1}) + \varphi(x_{m+t}))\Delta + O(\Delta^3)$.

From this it follows that

$$\sum_{x_k \in [-T\Delta, T\Delta)} \varphi(x_k)\Delta = \int_{-T\Delta}^{T\Delta} \varphi(x)\,dx + O(\Delta^3 T), \quad (3)$$

thus proving the lemma.

Next, suppose x_0 is not a multiple of Δ. Applying Lemma 1, as in the previous case, we arrive at (2). Let us consider a new scheme in which $\bar{x}_k = \bar{x}_0 + k\Delta$ and $\bar{x}_0 = [x_0/\Delta]\Delta = -m\Delta$. It is easy to verify that, since $\varphi(x)$ is even and its second derivative is bounded, we have, for any integer t,

$$\varphi(x_{m-t-1}) + \varphi(x_{m+t}) = \varphi(\bar{x}_{m-t-1}) + \varphi(\bar{x}_{m+t}) + O(\Delta^2).$$

Hence it follows that

$$\sum_{x_k \in [-T\Delta, T\Delta)} \varphi(x_k)\Delta = \sum_{\bar{x}_k \in [-T\Delta, T\Delta)} \varphi(\bar{x}_k)\Delta + O(\Delta^3 T).$$

When we apply (3) to the summation on the right, the assertion of the lemma follows in this case as well.

Lemma 3. *As* $\sigma \to \infty$,

$$\sum_{k=-\infty}^{\infty} N(k, a, \sigma^2) = 1 + O\left(\frac{\sqrt{\ln \sigma}}{\sigma^2}\right).$$

Proof. If in Lemma 2 we take $\varphi(x) = \frac{1}{\sqrt{2\pi}} e^{-x^2/2}$, $\Delta = 1/\sigma$, $x_k = (k-a)/\sigma$, and $T = [\sigma\sqrt{2 \ln \sigma}]$, we obtain the assertion of the lemma.

We estimate now the distances $\rho_i(n, p)$ for $i = 1, 2$. Let us define

$$c_1 = \frac{1}{2\sqrt{2\pi}} \int_{-\infty}^{\infty} |1 - u^2| e^{-u^2/2} du = \sqrt{\frac{2}{\pi e}},$$

$$c_2 = \frac{1}{6\sqrt{2\pi}} \int_{-\infty}^{\infty} |u^3 - 3u| e^{-u^2/2} du = \frac{1 + 4e^{-3/2}}{3\sqrt{2\pi}}.$$

Theorem 4. *If* $npq \to \infty$ *as* $p \to 0$ *and* $n \to \infty$ *then*

$$\rho_2(n, p) = \frac{c_2}{\sqrt{npq}}(1 + o(1)).$$

Theorem 5. *If $np \to \infty$ as $p \to 0$, then*

$$\rho_1(n, p) = c_1 p(1 + o(1)).$$

PROOF of Theorem 4. Let us define $v = (npq)^{1/7}$. Let us represent $\rho_2(n, p)$ as the sum $\rho_2(n, p) = A_1 + A_2$, where

$$A_1 = \sum_{|u| \leqslant v} |C_n^k p^k q^{n-k} - N(k, np, npq)|,$$

$$A_2 = \sum_{|u| > v} |C_n^k p^k q^{n-k} - N(k, np, npq)|,$$

$$u = \frac{k - np}{\sqrt{npq}}.$$

Note that, for any fixed m,

$$\int_v^\infty |u|^m e^{-u^2/2} du = O(v^{m-1} e^{-v^2/2}) \tag{4}$$

as $v \to \infty$.

Theorem 1 holds true in the domain $|u| \leqslant v$. Hence, applying Lemma 1 to replace summation with integration and using (4), we obtain for A_1 the estimate

$$A_1 = \frac{q - p}{6\sqrt{2\pi npq}} \int_{-\infty}^{\infty} |u^3 - 3u| e^{-u^2/2} du + O\left(\frac{1}{npq}\right). \tag{5}$$

Let us note that, for any fixed m,

$$\sum_{k=0}^{n} |u|^m C_n^k p^k q^{n-k} = O(1). \tag{6}$$

Using Chebyshev's inequality, we find that

$$\sum_{|u| > v} C_n^k p^k q^{n-k} \leqslant \frac{1}{v^7} \sum_{k=0}^{n} |u|^7 C_n^k p^k q^{n-k} = O\left(\frac{1}{npq}\right). \tag{7}$$

Using Lemma 1, we obtain the estimate

$$\sum_{|u|>v} N(k, np, npq) = O\left(\frac{1}{npq}\right). \tag{8}$$

It follows from (7) and (8) that $A_2 = O\left(\frac{1}{npq}\right)$; this estimate together with the estimate (5) leads to the assertion of Theorem 4.

PROOF of Theorem 5. Let us define $v=p^{-1/3}$. Let us represent $\rho_1(n, p)$ as the sum $\rho_1(n, p) = A_1 + A_2$, where

$$A_1 = \sum_{|u|\leqslant v} |C_n^k p^k q^{n-k} - \Pi(k, np)|,$$

$$A_2 = \sum_{|u|>v} |C_n^k p^k q^{n-k} - \Pi(k, np)|,$$

$$u = \frac{k-np}{\sqrt{np}}.$$

Note that, for any fixed m,

$$\sum_{k=0}^{\infty} |u|^m \Pi(k, np) = O(1). \tag{9}$$

Applying Theorem 2, we find that

$$A_1 = \frac{1}{2} p \sum_{|u|\leqslant v} |1-u^2| \Pi(k, np) + O\left(p^2 + \frac{p}{\sqrt{np}}\right). \tag{10}$$

Using Theorem 3 with $|u| \leqslant v^* = \min(p^{-1/3}, (np)^{1/7})$ and then using Lemma 1 to replace summation with integration, we see that

$$\sum_{|u|\leqslant v} |1-u^2| \Pi(k, np) = \sum_{|u|\leqslant v^*} |1-u^2| N(k, np, np) + o(1) =$$

$$= \frac{1}{\sqrt{2\pi}} \int_{-\infty}^{\infty} |1-u^2| e^{-u^2/2} du + o(1).$$

From this it follows that

$$A_1 = c_1 p(1+o(1)). \tag{11}$$

Taking into (6) and (9) and using Chebyshev's inequality, we find that

$$\sum_{|u|>v} C_n^k p^k q^{n-k} \leqslant \frac{1}{v^4} \sum_{k=0}^{n} u^4 C_n^k p^k q^{n-k} = O(p^{4/3}),$$

$$\sum_{|u|>v} \Pi(k, np) \leqslant \frac{1}{v^4} \sum_{k=0}^{\infty} u^4 \Pi(k, np) = O(p^{4/3}).$$

From this it follows that $A_2 = O(p^{4/3})$, which, together with the estimate given in (11), leads to the assertion of the theorem.

It is seen from Theorems 4 and 5 that the distance $\rho_1(n, p)$ to a Poisson distribution decreases as p decreases and that the distance $\rho_2(n, p)$ to a normal distribution increases as p decreases from 1/2 to 0. The point p^* at which a normal approximation and a Poisson approximation have the same accuracy in the sense of the distance with respect to variation is a root of the equation $\rho_1(n, p) = \rho_2(n, p)$. By Theorems 4 and 5, this equation is of the form

$$\frac{c_2}{\sqrt{npq}} = c_1 p(1 + o(1)),$$

and p^* is asymptotically equal to

$$p^* = \left(\frac{c_2^2}{c_1^2 n}\right)^{1/3} (1 + o(1)).$$

Of the two limit distributions, if $p < p^*$, a Poisson distribution yields the better approximation; if $p > p^*$, a normal distribution yields the better approximation. Since a binomial distribution is symmetric, the function $\min_{1 \leqslant i \leqslant 3} \rho_i(n, p)$ on the interval $0 \leqslant p \leqslant 1$

is symmetric with respect to $p=1/2$. Also this function has two equal maxima if $p=p^*$ and $p=1-p^*$. We have the worst approximation at p^*:

$$\sup_{0\leqslant p\leqslant 1}\min_i \rho_i(n,p)=\rho_1(n,p^*)=$$
$$=\rho_2(n,p^*)=\left(\frac{c_1c_2^2}{n}\right)^{1/3}(1+o(1)).$$

We investigate in a similar manner the rate of convergence of the distribution of $\mu_r(n, N)$ to the limit distribution. To refine limit Theorems 3.2 and 3.3, we shall need to estimate the remainder term in Theorem 3.1. We shall use the notation of Section 3.

Theorem 6. *If* $\alpha m\to\infty$ *and* $\frac{1}{\sqrt{\alpha m}}e^{z^2/2}\to 0$ *as* $m\to\infty$, *where* $z=\frac{l-m\alpha_r}{\sigma_r\sqrt{m}}$, *then, for any fixed* $r\geqslant 2$, *we have*

$$\mathbf{P}\{\zeta_m^{(r)}=l\}=\frac{1}{\sigma_r\sqrt{2\pi m}}e^{-z^2/2}\left(1+O\left(\frac{1}{\sqrt{\alpha m}}e^{z^2/2}\right)\right)$$

PROOF. As in proving Theorem 3.1, we express the difference

$$R_m=2\pi\left[\sigma_r\sqrt{m}\,\mathbf{P}\{\zeta_m^{(r)}=l\}-\frac{1}{\sqrt{2\pi}}e^{-z^2/2}\right] \tag{12}$$

as the sum of the four integrals mentioned above: $R_m=I_1+I_2+I_3+I_4$. As we have noted, for $\varphi_r(t)=\ln f_r^*(t)$,

$$\varphi_r(t)=-\frac{\sigma_r^2t^2}{2}(1+O(t)), \tag{13}$$

as $t\to 0$ since, for any mode of variation of α, the variable $|\sigma_r^{-2}\varphi'''(t)|$ is bounded in a neighborhood of the point $t=0$.

In the integrals I_1, I_2, and I_4, we choose $A=(\sigma_r\sqrt{m})^{1/4}$. Then it follows from (13) that, for $|t|\leqslant A$,

$$\left(f^*\left(\frac{t}{\sigma_r\sqrt{m}}\right)\right)^m = e^{-t^2/2}\left(1+O\left(\frac{|t|^3}{\sigma_r\sqrt{m}}\right)\right)$$

and that

$$|I_1| \leqslant \frac{c_1}{\sigma_r\sqrt{m}}\int_{-A}^{A}|t|^3 e^{-t^2/2}dt \leqslant \frac{C_1}{\sigma_r\sqrt{m}}$$

where c_1 and C_1 are positive constants. The integrals I_2, I_3, and I_4 are estimated in the same way as in the proof of Theorem 3.1. Noting that $A=(\sigma_r\sqrt{m})^{1/4}$, we find from the estimates obtained that the integrals I_2, I_3, and I_4 are variables of order $O\left(\frac{1}{\sigma_r\sqrt{m}}\right)$ as $m\to\infty$. Therefore, we infer that

$$R_m = O\left(\frac{1}{\sigma_r\sqrt{m}}\right) \tag{14}$$

as m, $\alpha m\to\infty$.

The assertion of Theorem 6 follows immediately from the estimate (14), the representation (12), and the fact that $\sigma_r^2=\alpha(1+o(1))$ both as $\alpha\to 0$ and as $\alpha\to\infty$.

We shall now refine Theorems 3.2 and 3.3 in the left- and right-hand intermediate r-domains.

Theorem 7. *Let r denote an integer in* $[2,\infty)$. *Suppose that* $p_r\to 0$ *and* $\frac{1+u_r^6}{Np_r}\to 0$ *as* $n\to\infty$ *and that* $\beta u_r^2\leqslant c<\infty$, *where*

$$u_r=\frac{k-Np_r}{\sigma_{rr}\sqrt{N}} \quad \text{and} \quad \beta=\frac{(\alpha-r)^2}{\alpha}p_r. \textit{ Then,}$$

$$\mathbf{P}\{\mu_r(n,N)=k\} = \frac{1}{\sigma_{rr}\sqrt{2\pi N}}e^{-u_r^2/2}\left(1+\frac{3u_r-u_r^3}{6\sqrt{Np_r}}+\right.$$
$$\left.+O\left(\frac{(1+|u_r|^3)\sqrt{\beta}}{\sqrt{Np_r}}+\frac{1+u_r^6}{Np_r}\right)\right). \tag{15}$$

PROOF. By Theorem 1,

$$C_N^k p_r^k (1-p_r)^{N-k} = \\ = \frac{1}{\sqrt{2\pi N p_r (1-p_r)}} e^{-u^2/2} \left(1 + \frac{(1-2p_r)(3u-u^3)}{6\sqrt{Np_r(1-p_r)}} + \right. \\ \left. + O\left(\frac{1+u^6}{Np_r(1-p_r)}\right)\right), \quad (16)$$

where $u = \frac{k - Np_r}{\sqrt{Np_r(1-p_r)}}$. Since $\sigma_{rr}^2 = p_r(1-p_r-\beta)$ and

$$u = u_r \sqrt{\frac{1-p_r-\beta}{1-p_r}} = u_r(1+O(\beta)),$$

it follows from (16) that

$$C_N^k p_r^k (1-p_r)^{N-k} = \frac{1}{\sqrt{2\pi N p_r (1-p_r)}} e^{-u^2/2} \left(1 + \frac{3u_r - u_r^3}{6\sqrt{Np_r}} + \right. \\ \left. + O\left(\frac{|u_r| + |u_r|^3}{\sqrt{Np_r}} \beta + \frac{1+u_r^6}{Np_r}\right)\right). \quad (17)$$

Assuming in Theorem 6 that $m=N-k$ and $l=n-kr$ and substituting the explicit expressions for α_r and σ_r^2, we find that

$$\frac{(l-m\alpha_r)^2}{2m\sigma_r^2} = \frac{\beta u_r^2}{2(1-p_r)\left(1 - \frac{u_r\sigma_{rr}}{(1-p_r)\sqrt{N}}\right)} = \\ = \frac{\beta u_r^2}{2(1-p_r)} + O\left(\frac{|u_r|^3 \beta p_r}{\sqrt{Np_r}}\right). \quad (18)$$

Taking into account the condition $u_r^2 \beta \leqslant c < \infty$, we obtain

$$\mathbf{P}\{\zeta_{N-k}^{(r)}=n-rk\}=$$

$$=\frac{1}{\sigma_r\sqrt{2\pi N(1-p_r)}}e^{-\frac{\beta u_r^2}{2(1-p_r)}}\left(1+O\left(\frac{1}{\sqrt{\alpha N}}+\frac{|u_r|+|u_r|^3}{\sqrt{Np_r}}\beta\right)\right). \tag{19}$$

Finally,

$$\mathbf{P}\{\zeta_N=n\}=\frac{n^n e^{-n}}{n!}=\frac{1}{\sqrt{2\pi n}}\left(1+O\left(\frac{1}{\sqrt{\alpha N}}\right)\right). \tag{20}$$

Substituting (17), (19), and (20) into (3.2) and noting that $\frac{1}{\sqrt{\alpha N}}=O\left(\sqrt{\frac{\beta}{Np_r}}\right)$, we arrive at the assertion of the theorem

Theorem 8. *Let* r *denote an integer in* $[2,\infty)$. *Suppose that* $p_r\to 0$, $Np_r\to\infty$, *and* $\beta u^2\to 0$ *as* N, $n\to\infty$, *where* $u=\frac{k-Np_r}{\sqrt{Np_r}}$ *and* $\beta=\frac{(\alpha-r)^2}{\alpha}p_r$. *Then*

$$\mathbf{P}\{\mu_r(n,N)=k\}=\Pi(k,Np_r)\Big(1+\frac{1}{2}(1-u^2)\beta+$$

$$+O\left((1+u^2)p_r+(1+u^4)\beta^2+\frac{1}{\sqrt{\alpha N}}\right)\Big). \tag{21}$$

PROOF. By Theorem 2, we have

$$C_N^k p_r^k(1-p_r)^{N-k}=\Pi(k,Np_r)(1+O((1+u^2)p_r)). \tag{22}$$

From (18), (19), and the fact that $\sigma_r^2=\alpha(1-p_r-\beta)(1-p_r)^{-2}$, we find

$$\mathbf{P}\{\zeta_{N-k}^{(r)}=n-kr\}=\frac{1}{\sqrt{2\pi\alpha N}}\Big(1+\frac{1}{2}(1-u^2)\beta+$$

$$+O\left((1+u^4)\beta^2+\frac{1}{\sqrt{\alpha N}}\right)\Big). \tag{23}$$

Substituting (20), (22), and (23) into (3.2) we arrive at the assertion of the theorem.

Note that Theorem 8 yields the principal term of an asymptotic expansion if β approaches zero more slowly than $1/\sqrt{\alpha N}$.

Using the notation (1), we introduce the distances between the distribution of $\mu_r(n, N)$ and limit distributions:

$$\rho_{1r} = \sum_{k=0}^{\infty} |\mathbf{P}\{\mu_r(n, N) = k\} - \Pi(k, Np_r)|,$$

$$\rho_{2r} = \sum_{k=0}^{\infty} |\mathbf{P}\{\mu_r(n, N) = k\} - N(k, Np_r, N\sigma_{rr}^2)|.$$

Let us estimate the distances ρ_{1r} and ρ_{2r} in domains in which

$$n,\ Np_r \to \infty \text{ and } p_r \to 0.$$

Theorem 9. *Let* $r \geqslant 2$ *be fixed. If* $Np_r \to \infty$ *and* $p_r \to 0$ *as* $N, n \to \infty$, *then*

$$\rho_{1r} = c_1 \frac{(\alpha - r)^2}{\alpha} p_r (1 + o(1)) + O\left(\frac{1}{\sqrt{\alpha N}}\right).$$

PROOF. Let us define $\beta = \frac{(\alpha - r)^2}{\alpha} p_r$ and $v = \beta^{-1/3}$. We represent ρ_{1r} as the sum $\rho_{1r} = A_1 + A_2$, where

$$A_1 = \sum_{|u| \leqslant v} |\mathbf{P}\{\mu_r(n, N) = k\} - \Pi(k, Np_r)|,$$

$$A_2 = \sum_{|u| > v} |\mathbf{P}\{\mu_r(n, N) = k\} - \Pi(k, Np_r)|, \qquad u = \frac{k - Np_r}{\sqrt{Np_r}}.$$

With the aid of Theorem 8, we find that

$$A_1 = \frac{1}{2} \beta \sum_{|u| \leqslant v} |1 - u^2| \, \Pi(k, Np_r) + O\left(\beta^2 + p_r + \frac{1}{\sqrt{\alpha N}}\right).$$

Just as in the proof of Theorem 5, we obtain

$$\sum_{|u|\leqslant v} |1-u^2|\,\Pi(k, Np_r) = \frac{1}{\sqrt{2\pi}}\int_{-\infty}^{\infty} |1-u^2|\, e^{-u^2/2}du + o(1).$$

From this it follows that

$$A_1 = c_1\beta(1+o(1)) + O\left(\frac{1}{\sqrt{\alpha N}}\right). \tag{24}$$

By direct calculation, we obtain

$$\mathbf{M}\left(\frac{\mu_r - Np_r}{\sqrt{Np_r}}\right)^4 = O(1).$$

Hence, applying Chebyshev's inequality, we infer that

$$\sum_{|u|>v} \mathbf{P}\{\mu_r(n, N) = k\} \leqslant \frac{1}{v^4}\mathbf{M}\left(\frac{\mu_r - Np_r}{\sqrt{Np_r}}\right)^4 = O(\beta^{4/3}) \tag{25}$$

and that

$$\sum_{|u|>v} \Pi(k, Np_r) \leqslant \frac{1}{v^4}\sum_{k=0}^{\infty} u^4\Pi(k, Np_r) = O(\beta^{4/3}).$$

From this it follows that $A_2 = O(\beta^{4/3})$, which together with the estimate (24), leads to the assertion of the theorem.

Theorem 10. *Let $r \geqslant 2$ be fixed. If $p_r \to 0$ and $Np_r \to \infty$ as $N, n \to \infty$, then*

$$\rho_{2r} = \frac{c_2}{\sqrt{Np_r}}(1+o(1)).$$

PROOF. We set $v_r = \sqrt{\ln Np_r - \ln\beta}$ and represent ρ_{2r} as the sum $\rho_{2r} = A_1 + A_2$, where

$$A_1 = \sum_{|u_r| \leqslant v_r} \left| \mathbf{P}\{\mu_r(n, N) = k\} - N\left(k, Np_r, N\sigma_{rr}^2\right) \right|,$$

$$A_2 = \sum_{|u_r| > v_r} \left| \mathbf{P}\{\mu_r(n, N) = k\} - N\left(k, Np_r, N\sigma_{rr}^2\right) \right|,$$

$$u_r = \frac{k - Np_r}{\sigma_{rr}\sqrt{N}}.$$

Theorem 7 holds true in a domain $|u_r| \leqslant v_r$. Hence, using Lemma 1 to replace summation with integration, we obtain for A_1 the estimate

$$A_1 = \frac{c_2}{\sqrt{Np_r}}(1 + o(1)). \tag{26}$$

Using Lemma 3, we have

$$\sum_{|u_r| > v_r} \mathbf{P}\{\mu_r(n, N) = k\} = 1 - \sum_{|u_r| \leqslant v_r} N\left(k, Np_r, N\sigma_{rr}^2\right) +$$

$$+ \sum_{|u_r| \leqslant v_r} \left(N\left(k, Np_r, N\sigma_{rr}^2\right) - \mathbf{P}\{\mu_r(n, N) = k\}\right) =$$

$$= \sum_{|u_r| \leqslant v_r} \left(N\left(k, Np_r, N\sigma_{rr}^2\right) - \mathbf{P}\{\mu_r(n, N) = k\}\right) +$$

$$+ \sum_{|u_r| > v_r} N\left(k, Np_r, N\sigma_{rr}^2\right) + O\left(\frac{\sqrt{\ln Np_r}}{Np_r}\right). \tag{27}$$

Applying Lemma 1, we find

$$\sum_{|u_r| > v_r} N\left(k, Np_r, N\sigma_{rr}^2\right) = \frac{2}{\sqrt{2\pi}} \int_{v_r}^{\infty} e^{-u^2/2} du +$$

$$+ O\left(\frac{1}{\sigma_{rr}\sqrt{N}} e^{-v_r^2/2}\right) = O\left(\sqrt{\frac{\beta}{Np_r}}\right). \tag{28}$$

Using Theorem 7 and the fact that

$$\int_{-\infty}^{\infty} (u^3 - 3u)\, e^{-u^2/2} du = 0,$$

we obtain

$$\sum_{|u_r| \leqslant v_r} \left(N\left(k, Np_r, N\sigma_{rr}^2\right) - \mathbf{P}\{\mu_r(n, N) = k\}\right) =$$

$$= O\left(\frac{1}{\sqrt{Np_r}} e^{-v_r^2/2} + \sqrt{\frac{\beta}{Np_r}} + \frac{1}{Np_r}\right) = O\left(\sqrt{\frac{\beta}{Np_r}}\right). \quad (29)$$

From (27), (28), and (29) we get the estimate

$$A_2 = O\left(\sqrt{\frac{\beta}{Np_r}} + \sqrt{\frac{\ln Np_r}{Np_r}}\right),$$

which, together with (26), leads to the assertion of the theorem.

With the aid of Theorems 9 and 10, we can show graphically the dependence of the distances ρ_{1r} and ρ_{2r} on the parameter α.

In this figure, the segment (A_0, B_0) corresponds to the left-hand intermediate domain and the segment (B_∞, A_∞) corresponds to the right-hand intermediate domain. The segments (A_0, C_0) and (C_∞, A_∞) correspond to the domains in which Theorem 9 does not yield the correct value of the principal term of the distance ρ_{1r}. The estimate indicated in Theorem 9 of the distance $\rho_{1r} = O(1/\sqrt{\alpha N})$ in these domains enables us to conclude that

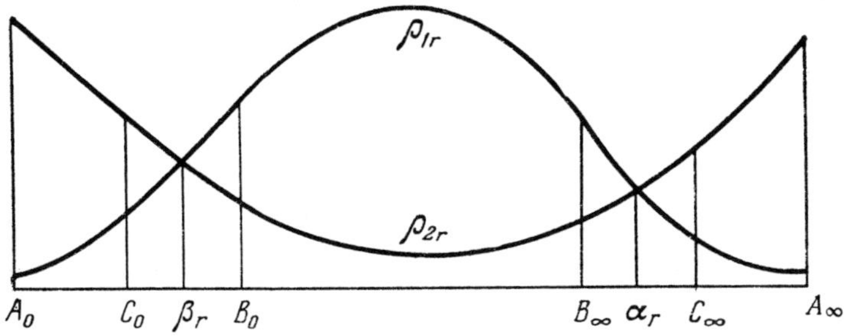

$\rho_{1r}<\rho_{2r}$ since $\rho_{2r}=\frac{c_2}{\sqrt{Np_r}}(1+o(1))$ here. In the domains (C_0, B_0) and (B_∞, C_∞), the principal terms of the distances ρ_{1r} and ρ_{2r} are monotone functions of the parameter α. The points α_r and β_r at which the distances to the two limit distributions coincide are roots of the equation $\rho_{1r}=\rho_{2r}$ and they are asymptotically equal to

$$\alpha_r=\frac{1}{3}\ln n+\left(r+\frac{1}{3}\right)\ln\ln n-\left(r+\frac{1}{3}\right)\ln 3-$$

$$-\ln r!+\frac{2}{3}\ln\frac{c_1}{c_2}+o(1), \qquad (30)$$

$$\beta_r=((r-1)!)^{\frac{1}{r-1}}\left(\frac{c_2^2}{rnc_1^2}\right)^{\frac{1}{3(r-1)}}(1+o(1)).$$

We have the worst approximation at the point α_r at which the distance to both limit distributions is asymptotically equal to

$$\sup_\alpha \min_i \rho_{ir}=\left(\frac{c_1c_2^2}{9n}\right)^{1/3}(\ln n+(3r+1)\ln\ln n)^{2/3}(1+o(1)),$$

where the supremum is over the left-hand and the right-hand intermediate domains. To prove this assertion as well as formulas (30), it suffices to substitute the values of α_r and β_r into the expressions for ρ_{1r} and ρ_{2r} given by Theorems 9 and 10.

Thus the behavior of the distributions of $\mu_r(n, N)$ is very similar to that of a binomial distribution. The absence of symmetry in the distributions of $\mu_r(n, N)$ is due to the fact that the principal term of the distance ρ_{1r} is equal to $c_1\frac{(\alpha-r)^2}{\alpha}p_r$ rather than c_1p_r, as is the case with the corresponding distance for a binomial distribution.

§6. Limit Distributions of Maximal and Minimal Occupancy of Boxes

Let us consider an equiprobable scheme of allocation of n particles in N cells. For i, $i=1, \ldots, N$, let η_i denote the number of particles in the ith cell. Arranging $\eta_1, \ldots, \eta_N$ in nondecreasing order, we construct the order sequence $\eta_{(1)}, \ldots \eta_{(N)}$. In the present section, we shall study the asymptotic behavior of the extreme terms of this sequence as $n, N \to \infty$. The forms of limit distributions of the terms of the order sequence as $n, N \to \infty$ depend on the behavior of the parameter $\frac{n}{N \ln N}$. If $\frac{n}{N \ln N} \to 0$ as $n, N \to \infty$, the distribution of each term of the series is concentrated asymptotically at one or two points. As the parameter $n/N \ln N$ increases, the limit distributions become more and more diffuse. If $n/N \ln N \to x$, $x > 0$ the distribution concentrated at countably many points is the limit distribution for $\eta_{(N-m+1)}$ for any fixed m. The spread of this distribution increases as x increases. The general behavior of the terms of the order sequence $\eta_{(m)}$ is similar to that of the sequence $\eta_{(1)}, \ldots, \eta_{(N)}$ described above, for any fixed m. The value $x = 1$ is critical. If $n/N \ln N \to x$, then for $x \leqslant 1$ the limit distributions are concentrated at one or two points; for $x > 1$, they are concentrated at countably many points. If $n/N \ln N \to \infty$, the extreme terms of the order sequence have in the limit the distributions with density of the double exponential type if these terms are properly centralized and standardized.

Distributions of the terms of an order sequence are related to distributions of random variables $\mu_r(n, N)$ in a simple manner; for example,

$$\mathbf{P}\{\eta_{(N)} \leqslant k\} = \mathbf{P}\{\mu_k + \mu_{k+1} + \ldots + \mu_n = 0\}.$$

Hence, in investigating the order sequence, the asymptotic properties of random variables of the form $\mu_k + \ldots + \mu_n$ can be used. In the present section, however, we take another approach, based on the representation (3.1) of the joint distribution of the random variables $\eta_1, \ldots, \eta_N$ and using independent random variables $\xi_1, \ldots, \xi_N$ distributed according to a Poisson law. In taking this approach, we investigate the minimal term and the maximal term of the order sequence in almost the same way. Hence, we shall first investigate in detail the behavior of the maximal term $\eta_{(N)}$

and then discuss briefly the results concerning $\eta_{(1)}$.

As usual, we define

$$\alpha = \frac{n}{N}, \ p_k = \frac{\alpha^k}{k!} e^{-\alpha}, \ k = 0, 1, \ldots.$$

The limit behavior $\eta_{(N)}$ is described by the following three theorems:

Theorem 1. *If $\alpha/\ln N \to 0$ as $n, N \to \infty$ and $r = r(\alpha, N)$ is chosen so that $r > \alpha$ and $Np_r \to \lambda$ where λ is a positive constant, then*

$$\mathbf{P}\{\eta_{(N)} = r - 1\} \to e^{-\lambda}, \qquad \mathbf{P}\{\eta_{(N)} = r\} \to 1 - e^{-\lambda}.$$

Theorem 2. *If $\alpha/\ln N \to x$ as $n, N \to \infty$ and $r = r(\alpha, N)$ is chosen so that $Np_r \to \lambda$, where x and λ are positive constants, then*

$$\mathbf{P}\{\eta_{(N)} \leqslant r + k\} \to \exp\left\{-\frac{\lambda\gamma^{k+1}}{1-\gamma}\right\},$$

where γ is the root of the equation

$$\gamma + x(\ln \gamma - \gamma + 1) = 0 \tag{1}$$

in an interval $0 < \gamma < 1$.

Theorem 3. *If $\alpha/\ln N \to 0$ as $n, N \to \infty$, then*

$$\mathbf{P}\left\{\frac{\eta_{(N)} - \alpha - \alpha u\left(\frac{1}{\alpha}\left(\ln N - \frac{\ln \ln n}{2}\right)\right)}{\sqrt{\alpha/2 \ln N}} + \frac{\ln 4\pi}{2} \leqslant z\right\} \to e^{-e^{-z}},$$

where $u(w)$ is a positive function defined on the interval $0 < w < \infty$ by the equation

$$-u + (1+u)\ln(1+u) = w. \tag{2}$$

To prove these theorems, we need some auxiliary results. Lemma 1 (on the relation between the distribution of $\eta_{(N)}$ and the Poisson distribution) and Lemma 7 are most crucial. From Lemmas 1, 7, and 8, it follows that the limit distribution of the maximum in the equiprobable scheme coincides with the limit distribution of the maximum of N independent random variables having a Poisson distribution with parameter α. To find the limit distribution of $\eta_{(N)}$, it suffices to consider quantiles of the Poisson distribution. Lemmas 2, 3, and 6 deal with this problem. In Lemma 5 a solution of Equation (2) is expressed as a series.

Let $\xi_1, \ldots, \xi_N$ be independent uniformly distributed random variables having a Poisson distribution with parameter α. We introduce independent random variables $\xi_1^{(A)}, \ldots, \xi_N^{(A)}$ for which

$$\mathbf{P}\{\xi_i^{(A)} = k\} = \mathbf{P}\{\xi_i = k \mid \xi_i \notin A\}, \quad i = 1, \ldots, N, \quad k \in \bar{A},$$

where A is a set of nonnegative integers and $\bar{A}$ is its complement in the set of all nonnegative integers.

Lemma 1.

$$\mathbf{P}\{\eta_{(N)} \leqslant r\} = (1 - \mathbf{P}\{\xi_1 > r\})^N \frac{\mathbf{P}\left\{\xi_1^{(A_r)} + \ldots + \xi_N^{(A_r)} = n\right\}}{\mathbf{P}\{\xi_1 + \ldots + \xi_N = n\}},$$

where A_r is the set of integers larger than r.

PROOF. By virtue of (3.1),

$$\mathbf{P}\{\eta_{(N)} \leqslant r\} = \mathbf{P}\{\xi_1 \leqslant r, \ldots, \xi_N \leqslant r \mid \xi_1 + \ldots + \xi_N = n\} =$$
$$= (\mathbf{P}\{\xi_1 \leqslant r\})^N \frac{\mathbf{P}\{\xi_1 + \ldots + \xi_N = n \mid \xi_1 \leqslant r, \ldots, \xi_N \leqslant r\}}{\mathbf{P}\{\xi_1 + \ldots + \xi_N = n\}} =$$
$$= (1 - \mathbf{P}\{\xi_1 > r\})^N \frac{\mathbf{P}\left\{\xi_1^{(A_r)} + \ldots + \xi_N^{(A_r)} = n\right\}}{\mathbf{P}\{\xi_1 + \ldots + \xi_N = n\}}.$$

Let us examine the behavior of the tails of the Poisson distribution $P_r = P_r(\alpha) = \mathbf{P}\{\xi_1 > r\}$ for different ranges of α and r.

Lemma 2. *If $\alpha/\ln N \to 0$ as $N \to \infty$ and $r = r(\alpha, N)$ is* *chosen so that* $r > \alpha$ *and* $Np_r \to \lambda$, *where* λ *is a positive constant, then*

$$\frac{\alpha}{r} \to 0,\ NP_r \to 0,\ NP_{r-1} \to \lambda,\ NP_{r-2} \to \infty.$$

PROOF. Taking the logarithm in $Np_r \to \lambda$ and multiplying both sides of the resulting relation by $\alpha/r \ln N$, we find that

$$\frac{\alpha}{r} + \frac{\alpha \ln \alpha}{\ln N} - \frac{\alpha^2}{r \ln N} - \frac{\alpha \ln r!}{r \ln N} = o(1). \tag{3}$$

If r is bounded, then, taking into account the relations $\alpha/\ln N \to 0$ and $\alpha/r < 1$, we obtain from (3)

$$\frac{\alpha}{r} + \frac{\alpha \ln \alpha}{\ln N} = o(1). \tag{4}$$

From this it follows that $\alpha \to 0$. In fact, it is impossible to have the inequality $\alpha > 1$ satisfied for an infinite sequence of values of α since in this case both terms in the left side of (4) are nonnegative and therefore must approach zero. On the other hand, for $0 < \alpha \leqslant 1$, the function $\alpha \ln \alpha$ is bounded. Therefore, it follows from (4) that $\alpha/r = o(1)$.

Applying Stirling's formula for the expansion $\ln r!$ we find that

$$\frac{\alpha}{r} + \frac{\alpha}{\ln N} \ln \frac{\alpha}{r} = o(1) \tag{5}$$

as $r \to \infty$.

Assuming that $\alpha/r \geqslant \delta > 0$ for an infinite sequence of values, we arrive at a contradiction because in this case $\dfrac{\alpha}{\ln N} \ln \dfrac{\alpha}{r} \to 0$ and it follows from (5) that $\alpha/r = o(1)$.

Thus, $\alpha/r \to 0$ under the conditions of Lemma 2. Hence, we infer that

$$NP_r = N \sum_{k=r+1}^{\infty} p_k = Np_r \sum_{k=r+1}^{\infty} \frac{\alpha^{k-r} r!}{k!},$$

$$NP_{r-1} = N \sum_{k=r}^{\infty} p_k = Np_r \left(1 + \sum_{k=r+1}^{\infty} \frac{\alpha^{k-r} r!}{k!}\right).$$

From this it follows that $NP_r \to 0$ and $NP_{r-1} \to \lambda$ since $Np_r \to \lambda$ and

$$\sum_{k=r+1}^{\infty} \frac{\alpha^{k-r} r!}{k!} \leqslant \sum_{k=1}^{\infty} \left(\frac{\alpha}{r}\right)^k = \frac{\alpha}{r}\left(1 - \frac{\alpha}{r}\right)^{-1} \to 0.$$

Furthermore, $NP_{r-2} \geqslant Np_{r-1} = Np_r \dfrac{r}{\alpha} \to \infty$.

Lemma 3. *If* $\alpha/\ln N \to x$ *as* $N \to \infty$ *and* $r = r(\alpha, N)$ *is chosen so that* $r > \alpha$ *and* $Np_r \to \lambda$, *where* x *and* λ *are positive constants, then*

$$\frac{\alpha}{r} \to \gamma \quad \text{and} \quad NP_{r+k} \to \frac{\lambda \gamma^{k+1}}{1-\gamma},$$

where k *is any fixed integer and* γ *is the root of equation* (1) *in the interval* $0 < \gamma < 1$.

PROOF. Taking the logarithm in $Np_r \to \lambda$ and multiplying both sides of the relation thus obtained by $\alpha/r \ln N$, we find that

$$\frac{\alpha}{r} + \frac{\alpha}{\ln N} \ln \frac{\alpha}{r} - \frac{\alpha^2}{r \ln N} + \frac{\alpha}{\ln N} = o(1). \tag{6}$$

Since $\alpha/r < 1$ and the solution of equation (1) is unique in the interval $0 < \gamma < 1$ and depends continuously on x, it follows from (6) that the limit α/r exists and satisfies equation (1). Also, for any fixed k,

$$NP_{r+k} = Np_r \sum_{l=r+k+1}^{\infty} \frac{p_l}{p_r}. \tag{7}$$

As established above, $\alpha/r \to \gamma$, hence $p_{r+s}/p_r \to \gamma^s$ for fixed s. Taking the limit in (7) under the summation sign, we obtain

$$NP_{r+k} \to \lambda \sum_{l=r+k+1}^{\infty} \gamma^{l-r} = \frac{\lambda\gamma^{k+1}}{1-\gamma}.$$

Lemma 4. *If* $\alpha \to \infty$ *and* $r \to \infty$ *so that* $x_\alpha = \frac{r-\alpha}{\sqrt{\alpha}} = o(\sqrt{\alpha})$ *and* $x_\alpha \geqslant 0$, *then*

$$P_r = (1 - G(x_\alpha)) \exp\left\{\frac{x_\alpha^3}{\sqrt{\alpha}} \lambda\left(\frac{x_\alpha}{\sqrt{\alpha}}\right)\right\}\left(1 + O\left(\frac{1+x_\alpha}{\sqrt{\alpha}}\right)\right),$$

$$p_r = \frac{1}{\sqrt{2\pi\alpha}} \exp\left\{-\frac{x_\alpha^2}{2} + \frac{x_\alpha^3}{\sqrt{\alpha}} \lambda\left(\frac{x_\alpha}{\sqrt{\alpha}}\right)\right\}\left(1 + O\left(\frac{1+x_\alpha}{\sqrt{\alpha}}\right)\right),$$

where $\lambda(z) = \frac{1}{z^3}\left(z + \frac{z^2}{2} - (1+z)\ln(1+z)\right)$ *and*

$$G(z) = \frac{1}{\sqrt{2\pi}} \int_{-\infty}^{z} e^{-x^2/2}\, dx.$$

PROOF. If $\alpha \to \infty$ through integral values, ξ_1 can be represented as the sum of independent uniformly distributed random variables $X_1, \ldots, X_\alpha$ having a Poisson distribution with parameter 1. We arrive at the assertion of Lemma 4 by applying directly the integral theorem and the local theorem on large deviations for the sum $X_1 + \ldots + X_\alpha$. Cramér's series $\lambda(\tau)$ involved in these theorems is defined by

$$\lambda(\tau) = \frac{1}{\tau^3}\left(\frac{\tau^2}{2} + K(z_0) - \tau z_0\right),$$

where $K(z)$ is the principal branch of the function $\ln M(z)$

(where, in turn, $M(z) = Me^{z(X_1-1)} = e^{e^z - z - 1}$) and z_0 is the root of the equation $K'(z) = \tau$. It is easy to verify that $\lambda(\tau)$ is of the form indicated in the lemma.

The shift to arbitrary convergence of α to infinity involves no difficulties since $p_r = p_r(\alpha)$ and $P_r = P_r(\alpha)$ for $r \geqslant \alpha$ are monotone with respect to α and

$$p_r(\alpha + 1) = p_r(\alpha)\left(1 + O\left(\frac{x_\alpha}{\sqrt{\alpha}}\right)\right),$$

$$P_r(\alpha + 1) = P_r(\alpha)\left(1 + O\left(\frac{x_\alpha}{\sqrt{\alpha}}\right)\right),$$

since, for $x_\alpha > 0$ and $x_\alpha = o(\sqrt{\alpha})$, we have the estimate

$$p_r(\alpha + 1) - p_r(\alpha) = p_r(\alpha)\left(\frac{1}{e}\left(1 + \frac{1}{\alpha}\right)^r - 1\right) \leqslant$$

$$\leqslant p_r(\alpha)\left(e^{x_\alpha/\sqrt{\alpha}} - 1\right) = O\left(p_r(\alpha)\frac{x_\alpha}{\sqrt{\alpha}}\right).$$

Consider next solutions of equation (2). The function $w = -u + (1+u)\ln(1+u)$ is strictly decreasing in the interval $[-1, 0]$ and strictly increasing in the interval $[0, \infty)$. Hence, equation (2) has a positive solution $u(w)$ and a negative solution $v(w)$.

For $|u| < 1$, we have

$$-u + (1 + u)\ln(1 + u) = \sum_{k=2}^{\infty} \frac{(-u)^k}{k(k-1)},$$

so that, by applying standard formulas, we obtain the following result:

Lemma 5. *The equation*

$$-u + (1+u)\ln(1+u) = w$$

has a nonnegative solution $u(w)$ *in the interval* $0 \leqslant w < \infty$ *and has a nonpositive solution* $v(w)$ *in the interval* $0 \leqslant w < 1$. *In some neighborhood of zero, these solutions can be expressed by the convergent series*

$$u(w) = \sqrt{2w} + \sum_{l=2}^{\infty} (-1)^{l-1} d_l \left(\sqrt{2w}\right)^l, \tag{8}$$

$$v(w) = -\sqrt{2w} - \sum_{l=2}^{\infty} d_l \left(\sqrt{2w}\right)^l. \tag{9}$$

In these formulas,

$$d_{l+1} = \frac{1}{l+1} \sum \frac{(-1)^k (l+1)(l+3)\cdots(l+2k-1)}{k_1! \ldots k_l! \, (2\cdot 3)^{k_1} \ldots ((l+1)(l+2))^{k_l}}, \quad l = 1, 2, \ldots,$$

where $k = k_1 + \ldots + k_l$ *and where the summation is over all nonnegative integers* $k_1, \ldots, k_l$ *such that* $k_1 + 2k_2 + \ldots + lk_l = l$.

It is easy to verify that

$$u(w) = \sqrt{2w} + \frac{1}{3} w - \frac{\sqrt{2}}{36} w^{3/2} + \ldots,$$

$$v(w) = -\sqrt{2w} + \frac{1}{3} w + \frac{\sqrt{2}}{36} w^{3/2} + \ldots .$$

Lemma 6. *If* $\alpha/\ln N \to \infty$ *as* $n, N, \to \infty$ *and if*

$$r = \alpha + \alpha u \left(\frac{\ln N - \frac{1}{2} \ln \ln N}{\alpha} \right) + \\ + \sqrt{\frac{\alpha}{2 \ln N}} \left(z - \frac{1}{2} \ln 4\pi + o(1) \right),$$

then

$$NP_r = e^{-z} + o(1), \qquad p_r = O\left(\frac{1}{N} \sqrt{\frac{\ln N}{\alpha}} \right).$$

PROOF. It follows from (8) that

$$u\left(\frac{\ln N-\frac{1}{2}\ln\ln N}{\alpha}\right)=\sqrt{\frac{2\ln N}{\alpha}}(1+o(1)).$$

Hence, for the choice of r indicated in the lemma, we obtain

$$x_\alpha=\frac{r-\alpha}{\sqrt{\alpha}}=\sqrt{2\ln N}(1+o(1))=o(\sqrt{\alpha}), \qquad (10)$$

and Lemma 4 can be used in estimating P_r. Applying Lemma 4 and the relation

$$1-G(x_\alpha)=\frac{1}{\sqrt{2\pi}x_\alpha}e^{-x_\alpha^2/2}(1+o(1)\text{ as } x_\alpha\to\infty, \qquad (11)$$

we get

$$NP_r=\frac{N}{\sqrt{2\pi}x_\alpha}\exp\left\{\alpha\left[\frac{x_\alpha}{\sqrt{\alpha}}-\left(1+\frac{x_\alpha}{\sqrt{\alpha}}\right)\ln\left(1+\frac{x_\alpha}{\sqrt{\alpha}}\right)\right]\right\}\times$$
$$\times(1+o(1)). \qquad (12)$$

Denote for brevity

$$u=u\left(\frac{\ln N-\frac{1}{2}\ln\ln N}{\alpha}\right),$$
$$\Delta=\frac{r-\alpha}{\alpha}-u=\frac{1}{\alpha}\sqrt{\frac{\alpha}{2\ln N}}\left(z-\frac{1}{2}\ln 4\pi+o(1)\right).$$

In this notation, we have $x_\alpha/\sqrt{\alpha}=u+\Delta$. Taking the logarithm in (12) and noting that, by virtue of (10),

$$\ln x_\alpha=\frac{1}{2}\ln\ln N+\frac{1}{2}\ln 2+o(1),$$

we obtain

$$\ln NP_r = \ln N - \frac{1}{2}\ln\ln N - \frac{1}{2}\ln 4\pi +$$
$$+ \alpha[u - (1+u)\ln(1+u)] + \alpha\Big[\Delta - \Delta\ln(1+u) -$$
$$- (1+u)\ln\left(1+\frac{\Delta}{1+u}\right) - \Delta\ln\left(1+\frac{\Delta}{1+u}\right)\Big] + o(1). \quad (13)$$

The variable u is chosen so that

$$\alpha[u - (1+u)\ln(1+u)] = -\ln N + \frac{1}{2}\ln\ln N.$$

Furthermore, since $\alpha\Delta^2 \to 0$ and $\alpha\Delta u^2 \to 0$, we have

$$\alpha\Big[\Delta - \Delta\ln(1+u) - (1+u)\ln\left(1+\frac{\Delta}{1+u}\right) -$$
$$- \Delta\ln\left(1+\frac{\Delta}{1+u}\right)\Big] = -\alpha\Delta u + o(1) = \frac{1}{2}\ln 4\pi - z + o(1).$$

Hence it follows from (13) that $\ln NP_r = -z + o(1)$.

The second assertion of the lemma follows from the fact that

$$p_r = \frac{x_\alpha}{\sqrt{\alpha}} P_r(1 + o(1)),$$

which in turn follows from Lemma 4 and equation (10). We estimate now the ratio of probabilities in Lemma 1. Let us define

$$\zeta_N^{(A_r)} = \xi_1^{(A_r)} + \ldots + \xi_N^{(A_r)}, \qquad m_r = \mathbf{M}\xi_1^{(A_r)}, \qquad \sigma_r^2 = \mathbf{D}\xi_1^{(A_r)}.$$

It can be easily calculated that

$$m_r = \alpha\left(1 - \frac{p_r}{1-P_r}\right), \quad \sigma_r^2 = \alpha\left(1 - \frac{p_r}{1-P_r} - \frac{(\alpha-r)p_r}{1-P_r} - \frac{\alpha p_r^2}{(1-P_r)^2}\right).$$

Let us prove that a local theorem on convergence to a normal distribution holds true for $\zeta_N^{(A_r)}$.

Lemma 7. *If* $r = r(\alpha, N)$ *is chosen so that* $r > \alpha$ *and* $NP_r \to \lambda$ *as* $n, N \to \infty$, *where* λ *is a positive constant, then*

$$\mathbf{P}\left\{\zeta_N^{(A_r)} = k\right\} = \frac{1}{\sigma_r \sqrt{2\pi N}} \exp\left\{-\frac{(k - Nm_r)^2}{2N\sigma_r^2}\right\}(1 + o(1))$$

uniformly with respect to $(k - Nm_r)/\sigma_r\sqrt{N}$ *in any finite interval.*

PROOF. Following the classical proof of a local limit theorem, we express the probability $\mathbf{P}\{\zeta_N^{(A_r)} = k\}$ in the form of an integral:

$$\mathbf{P}\left\{\zeta_N^{(A_r)} = k\right\} = \frac{1}{2\pi\sigma_r\sqrt{N}} \int_{-\pi\sigma_r\sqrt{N}}^{\pi\sigma_r\sqrt{N}} e^{-ixz}\left[f_r^*\left(\frac{x}{\sigma_r\sqrt{N}}\right)\right]^N dx,$$

where

$$z = \frac{k - Nm_r}{\sigma_r\sqrt{N}}, \qquad f_r^*(t) = e^{-itm_r} f_r(t),$$

$$f_r(t) = \mathbf{M}\exp\left(it\xi_1^{(A_r)}\right) = \frac{1}{1 - P_r}\sum_{k=0}^{r} p_k e^{itk}.$$

Since

$$\frac{1}{\sqrt{2\pi}} e^{-z^2/2} = \frac{1}{2\pi}\int_{-\infty}^{\infty} \exp\left(-ixz - \frac{x^2}{2}\right) dx,$$

the difference

$$R_N = 2\pi\left(\sigma_r\sqrt{N}\,\mathbf{P}\left\{\zeta_N^{(A_r)} = k\right\} - \frac{1}{\sqrt{2\pi}} e^{-z^2/2}\right)$$

can be represented as the sum of three integrals: $R_N = I_1 + I_2 + I_3$, where

$$I_1 = \int_{-A}^{A} e^{-izx} \left(\left(f_r^* \left(\frac{x}{\sigma_r \sqrt{N}} \right) \right)^N - e^{-x^2/2} \right) dx,$$

$$I_2 = \int_{A<|x|\leqslant \pi\sigma_r\sqrt{N}} e^{-izx} \left(f_r^* \left(\frac{x}{\sigma_r \sqrt{N}} \right) \right)^N dx,$$

$$I_3 = - \int_{A<|x|} e^{-izx - \frac{x^2}{2}} dx.$$

We shall choose the constant A later on.

The major difficulty is in estimating the integral I_2. We divide its integration domain into four parts:

$$S_1 = \{x\colon A < |x| \leqslant B\sqrt{N},\ |x| \leqslant \varepsilon\sigma_r\sqrt{N}\},$$
$$S_2 = \{x\colon B\sqrt{N} < |x| \leqslant CN,\ |x| \leqslant \varepsilon\sigma_r\sqrt{N}\},$$
$$S_3 = \{x\colon CN < |x|,\ |x| \leqslant \varepsilon\sigma_r\sqrt{N}\},$$
$$S_4 = \{x\colon \varepsilon\sigma_r\sqrt{N} < |x| \leqslant \pi\sigma_r\sqrt{N}\},$$

where ε, B, and C are constants. Note that the domains S_2 and S_3 may prove to be empty.

Since, for sufficiently small ε, we have, for $|t| \leqslant \varepsilon$,

$$\left| e^{\alpha(e^{it}-1)} \right| \leqslant e^{-\alpha t^2/4}.$$

Also, by the choice of r,

$$\left| \sum_{k=r+1}^{\infty} p_k e^{itk} \right| \leqslant P_r \leqslant \frac{a}{N},$$

where a is a constant, we have the following estimate for the characteristic function $f_r(t)$ for $|t| \leqslant \varepsilon$:

$$|f_r(t)| = \frac{1}{1-P_r}\left|e^{\alpha(e^{it}-1)} - \sum_{k=r+1}^{\infty} p_k e^{itk}\right| \leqslant$$
$$\leqslant \frac{1}{1-P_r}\left(e^{-\alpha t^2/4} + \frac{a}{N}\right).$$

Hence, noting that $\sigma_r^2 = \alpha(1+o(1))$, we obtain the chain of inequalities:

$$\left|\int_{S_1} e^{-izx}\left(f_r^*\left(\frac{x}{\sigma_r\sqrt{N}}\right)\right)^N dx\right| \leqslant c_1 \int_A^{B\sqrt{N}} \left(e^{-x^2/4N} + \frac{a}{N}\right)^N dx \leqslant$$
$$\leqslant c_1\left(1+\frac{ae^{B^2/4}}{N}\right)^N \int_A^{B\sqrt{N}} e^{-x^2/4}\,dx \leqslant C_1\int_A^{\infty} e^{-x^2/4}dx,$$

where c_1 and C_1 are constants. By choosing A sufficiently large, this part of the integral I_2 can be made arbitrarily small.

Let us apply Abel's transformation in estimating $\sum_{k=r+1}^{\infty} p_k e^{itk}$.

Let

$$A_m = \sum_{k=1}^{m} \alpha_k \beta_k,$$

and let $B_k = \beta_1 + \ldots + \beta_k$. Then A_m can be expressed as

$$A_m = \alpha_m B_m - \sum_{k=1}^{m-1} (\alpha_{k+1} - \alpha_k) B_k. \tag{14}$$

If the α_i do not increase and are positive and if the estimate $|B_k| \leqslant L$ holds for all k, then, as can easily be seen, it follows from (14) that

$$|A_m| \leqslant L\alpha_1. \tag{15}$$

Putting $\alpha_k = p_{r+k}$ and $\beta_k = e^{itk}$, we find that

$$A_m = \sum_{k=r+1}^{m+r} p_k e^{itk} \text{ and } |B_k| \leqslant \frac{2}{|1-e^{it}|}.$$

The values of p_{r+k} are positive and do not increase; hence, by virtue of (15), for any m we infer that

$$|A_m| \leqslant \frac{2p_{r+1}}{|1-e^{it}|},$$

and for an infinite sum we have the estimate

$$\left|\sum_{k=r+1}^{\infty} p_k e^{itk}\right| \leqslant \frac{2p_{r+1}}{|1-e^{it}|}.$$

From this it follows that there is a constant c such that, for $|t| \leqslant \varepsilon$,

$$\left|\sum_{k=r+1}^{\infty} p_k e^{itk}\right| \leqslant \frac{cp_{r+1}}{|t|},$$

and, for $|t| > \varepsilon$,

$$\left|\sum_{k=r+1}^{\infty} p_k e^{itk}\right| \leqslant cp_{r+1}. \tag{16}$$

Hence, for $|t| \leqslant \varepsilon$, we have

$$|f_r(t)| \leqslant \frac{1}{1-P_r}\left(e^{-\alpha t^2/4} + \frac{cp_{r+1}}{|t|}\right)$$

and for the integral over the domain S_2 we have the estimate

$$\left|\int_{S_2} e^{-izx}\left(f_r^*\left(\frac{x}{\sigma_r\sqrt{N}}\right)\right)^N dx\right| \leqslant$$

$$\leqslant \int_{B\sqrt{N}}^{CN} \frac{1}{(1-P_r)^N}\left(e^{-x^2/4N} + \frac{cp_{r+1}\sigma_r\sqrt{N}}{|x|}\right)^N dx \leqslant$$

$$\leqslant c_2 N\left(e^{-B^2/4} + \frac{cp_{r+1}\sigma_r}{B}\right)^N.$$

The right side of this inequality approaches zero as $N \to \infty$ since, by Lemmas 2, 3, and 6, the estimate

$$p_{r+1}\sigma_r = O\left(\frac{\sqrt{\ln N}}{N}\right). \tag{17}$$

holds for the choice of r indicated in Lemma 7.

Since we have $e^{-x^2/4N} \leqslant \frac{1}{2|x|}$ in the domain S_3 and since $\frac{cp_{r+1}\sigma_r\sqrt{N}}{|x|} \leqslant \frac{1}{2|x|}$ by virtue of (17), we have in this domain

$$\left|f_r\left(\frac{x}{\sigma_r\sqrt{N}}\right)\right| \leqslant \frac{1}{(1-P_r)|x|} \quad \text{and}$$

$$\left|\int\limits_{S_3} e^{-izx}\left(f_r^*\left(\frac{x}{\sigma_r\sqrt{N}}\right)\right)^N dx\right| \leqslant c_3 \int\limits_{CN}^{\infty} \frac{dx}{x^N} \to 0.$$

Finally, for $|t| > \varepsilon$, by virtue of (16) and Lemma 6, we arrive at the estimate

$$|f_r(t)| = \frac{1}{1-P_r}\left|e^{\alpha(e^{it}-1)} - \sum_{k=r+1}^{\infty} p_k e^{itk}\right| \leqslant$$
$$\leqslant \frac{1}{1-P_r}\left(e^{-\delta\alpha} + \frac{1}{\sqrt{\alpha N}}\right),$$

where $\delta > 0$. Hence,

$$\left|\int\limits_{S_4} e^{-izx}\left(f_r^*\left(\frac{x}{\sigma_r\sqrt{N}}\right)\right)^N dx\right| \leqslant c_4\sigma_r\sqrt{N}\left(e^{-\delta\alpha} + \frac{1}{\sqrt{\alpha N}}\right)^N \leqslant$$
$$\leqslant C_4\sqrt{\alpha N}\left(e^{-\delta\alpha} + \frac{1}{\sqrt{\alpha N}}\right)^N \to 0$$

as $n, N \to \infty$.

The integral I_1 approaches zero for any fixed A since the integral limit theorem on convergence to a normal distribution holds for $\zeta_N^{(A_r)}$. Finally, the integral I_3 can be made arbitrarily small by suitable choice of A.

Lemma 8. *If $r=r(\alpha, N)$ is chosen so that $r>\alpha$ and $NP_r \to \lambda$ as n, $N \to \infty$, where λ is a positive constant, then*

$$\frac{\mathbf{P}\left\{\zeta_N^{(A_r)} = n\right\}}{\mathbf{P}\left\{\xi_1 + \ldots + \xi_N = n\right\}} = 1 + o(1).$$

PROOF. According to the previous lemma,

$$\mathbf{P}\left\{\zeta_N^{(A_r)} = n\right\} = \frac{1}{\sigma_r\sqrt{2\pi N}} \exp\left\{-\frac{(n-Nm_r)^2}{2N\sigma_r^2}\right\}(1+o(1)). \quad (18)$$

Using explicit expressions for m_r and σ_r^2 as well as Lemmas 2, 3, and 6, we find that

$$\sigma_r^2 = \alpha(1+o(1)),$$

$$\frac{(n-Nm_r)^2}{2N\sigma_r^2} = \frac{1}{2N\sigma_r^2}\left(\frac{\alpha N p_r}{1-P_r}\right)^2 = O\left(\alpha N p_r^2\right) = O\left(\frac{\ln N}{N}\right).$$

Substituting these expressions into (18), we obtain

$$\mathbf{P}\left\{\zeta_N^{(A_r)} = n\right\} = \frac{1}{\sqrt{2\pi\alpha N}}(1+o(1)).$$

The assertion of the lemma now follows since we have the approximation

$$\mathbf{P}\left\{\xi_1 + \ldots + \xi_N = n\right\} = \frac{n^n e^{-n}}{n!} = \frac{1}{\sqrt{2\pi\alpha N}}(1+o(1))$$

for the expression in the denominator.

We shall now prove Theorems 1, 2, and 3. If the conditions of Theorem 1 are satisfied we have, by Lemma 2,

$$(1-P_{r-2})^N = o(1), \quad (1-P_{r-1})^N = e^{-\lambda} + o(1),$$
$$(1-P_r)^N = 1 + o(1). \tag{19}$$

If the conditions of Theorem 2 are satisfied, we have, by Lemma 3,

$$(1-P_{r+k})^N = \exp\{-\lambda\gamma^{k+1}(1-\gamma)^{-1}\} + o(1). \tag{20}$$

If the conditions of Theorem 3 are satisfied, we have by Lemma 6

$$(1-P_r)^N = e^{-e^{-z}} + o(1). \tag{21}$$

Using the representation of the maximum distribution function given in Lemma 1 and noting that, by Lemma 8, in this representation the ratio of probabilities is $1+o(1)$, we obtain the assertions of the theorems from (19), (20), and (21).

Note that if α tends to infinity sufficiently rapidly, the formulation of Theorem 3 can be simplified by dropping unnecessary terms in the representation (8) of the function $u(w)$. For example, if $\alpha^m(\ln N)^{-m-2} \to \infty$, it is sufficient to keep the first m terms of the expansion (8) instead of $u(w)$ since in this case

$$\frac{\alpha}{\sqrt{\alpha/2 \ln N}} \sum_{l=m+1}^{\infty} d_l \left(\frac{\ln N - \frac{1}{2}\ln\ln N}{\alpha}\right)^{l/2} = o(1).$$

Let us examine now the asymptotic behavior of the minimal term of the order sequence $\eta_{(1)}$. The following assertions hold.

Theorem 4. *If* $Ne^{-\alpha} \to \infty$ *as* $n, N \to \infty$, *then*

$$\mathbf{P}\{\eta_{(1)}=0\} \to 1.$$

Theorem 5. *If* $\alpha/\ln N \to 1$ *as* $n, N \to \infty$ *and* $r=r(\alpha, N)$ *is chosen so that* $r<\alpha$ *and* $Np_r \to \lambda$, *where* λ *is a positive constant, then*

$$\mathbf{P}\{\eta_{(1)}=r\} \to 1-e^{-\lambda},$$
$$\mathbf{P}\{\eta_{(1)}=r+1\} \to e^{-\lambda}.$$

Theorem 6. *If* $\frac{\alpha}{\ln N} \to x>1$ *as* $n,\ N \to \infty$ *and* $r=r(\alpha,\ N)$ *is chosen so that* $r<\alpha$ *and* $Np_r \to \lambda$, *where* λ *is a positive constant, then*

$$\mathbf{P}\{\eta_{(1)} \leqslant r+k\} \to 1-\exp\left\{-\frac{\lambda\beta^{k+1}}{\beta-1}\right\},$$

where β *is the root of equation* (1) *in the interval* $1<\beta<\infty$.

Theorem 7. *If* $\alpha/\ln N \to \infty$ *as* $n,\ N \to \infty$, *then*

$$\mathbf{P}\left\{\frac{\eta_{(1)}-\alpha-\alpha v\left(\left(\ln N-\frac{1}{2}\ln\ln N\right)/\alpha\right)}{\sqrt{\alpha/2\ln N}}-\frac{1}{2}\ln 4\pi \leqslant -z\right\}=$$
$$=1-e^{-e^{-z}}+o(1),$$

where $v(w)$ *is the negative function defined by equation* (2) *on the interval* $0 \leqslant w \leqslant 1$.

To prove Theorem 4, we take advantage of the relation

$$\mathbf{P}\{\eta_{(1)}=0\}=\mathbf{P}\{\mu_0(n,\ N)>0\},$$

where $\mu_0(n,\ N)$ is the number of empty boxes if n particles are allocated uniformly in N cells. By Theorems 1.3.2 and 1.3.3, the random variable $\mu_0(n,\ N)$ with $Ne^{-\alpha} \to \infty$ is asymptotically normal; hence $\mathbf{P}\{\mu_0(n,\ N)=0\} \to 0$, thus proving Theorem 4.

Theorems 5, 6, and 7 are proved in much the same way as the corresponding assertions for the maximal term of $\eta_{(N)}$. Instead of Lemma 1, we use

Lemma 9.

$$\mathbf{P}\{\eta_{(1)}=r\}=1-(1-\mathbf{P}\{\xi_1 \leqslant r\})^N \frac{\mathbf{P}\left\{\xi_1^{(\bar{A}_r)}+\ldots+\xi_N^{(\bar{A}_r)}=n\right\}}{\mathbf{P}\{\xi_1+\ldots+\xi_N=n\}}.$$

Just as when we were considering $\eta_{(N)}$, we can show here that if $r<\alpha$ is such that $N\mathbf{P}\{\xi_1\leqslant r\}\to\lambda$, as $n, N\to\infty$, where $0<\lambda<\infty$, then the local limit theorem on convergence to the normal distribution holds for $\xi_1^{(\bar{A}_r)}+\ldots+\xi_N^{(\bar{A}_r)}$ and the ratio of probabilities tends to 1 in Lemma 9. Hence, to study the asymptotic behavior of $\mathbf{P}\{\eta_{(1)}\leqslant r\}$, it suffices to estimate the left-hand tails of a Poisson distribution of $\mathbf{P}\{\xi_1\leqslant r\}$.

Lemma 10. *If* $\frac{\alpha}{\ln N}\to 1$ *as* $n, N\to\infty$ *and* $r=r(\alpha, N)$ *is chosen so that* $r<\alpha$ *and* $Np_r\to\lambda$, *where* λ *is a positive constant, then*

$$N\mathbf{P}\{\xi_1\leqslant r\}\to\lambda,\ N\mathbf{P}\{\xi_1\leqslant r-1\}\to 0,\ N\mathbf{P}\{\xi_1\leqslant r+1\}\to\infty.$$

Lemma 11. *If* $\frac{\alpha}{\ln N}\to x>1$ *as* $n, N\to\infty$ *and* $r=r(\alpha, N)$ *is chosen so that* $r<\alpha$ *and* $Np_r\to\lambda$, *where* λ *is a positive constant, then*

$$\frac{\alpha}{r}\to\beta \text{ and } N\mathbf{P}\{\xi_1\leqslant r+k\}\to\frac{\lambda\beta^{k+1}}{\beta-1},$$

where β *is the root of equation* (1) *in the interval* $1<\beta<\infty$.

Lemma 12. *If* $\frac{\alpha}{\ln N}\to\infty$ *as* $n, N\to\infty$ *and if*

$$r=\alpha+\alpha v\left(\frac{\ln N-\frac{1}{2}\ln\ln N}{\alpha}\right)-\\ -\sqrt{\frac{\alpha}{2\ln N}}\left(z-\frac{1}{2}\ln 4\pi+o(1)\right),$$

then

$$N\mathbf{P}\{\xi_1\leqslant r\}=e^{-z}+o(1).$$

The proofs of Lemmas 10, 11, and 12 are analogous to those

of Lemmas 2, 3, and 6, respectively. Note that (as follows from Lemmas 3, 6, 11, and 12) even if $\alpha \to \infty$, the left and the right quantiles of the Poisson distribution are not arranged asymptotically in a symmetric manner about the point α.

We present here a brief proof of Lemma 12. Using the theorem on large deviations, we find that, for

$$y_\alpha = -\frac{r-\alpha}{\sqrt{\alpha}} \geqslant 0, \quad y_\alpha = o(\sqrt{\alpha}),$$

we have

$$\mathbf{P}\{\xi_1 \leqslant r\} = $$
$$= G(-y_\alpha)\exp\left(-\frac{y_\alpha^3}{\sqrt{\alpha}}\lambda\left(-\frac{y_\alpha}{\sqrt{\alpha}}\right)\right)\left(1 + O\left(\frac{1+y_\alpha}{\sqrt{\alpha}}\right)\right),$$

where $\lambda(z)$ and $G(z)$ are defined in Lemma 4. As follows from (9),

$$v\left(\frac{\ln N - \frac{1}{2}\ln\ln N}{\alpha}\right) = -\sqrt{\frac{2\ln N}{\alpha}}(1 + o(1)).$$

Hence, for the choice of r given in the lemma, we have

$$y_\alpha = \sqrt{2\ln N}(1 + o(1)) = o(\sqrt{\alpha}).$$

Denoting for brevity

$$v = v\left(\frac{\ln N - \frac{1}{2}\ln\ln N}{\alpha}\right), \quad \Delta = \frac{1}{\alpha}\sqrt{\frac{\alpha}{2\ln N}}\left(z - \frac{1}{2}\ln 4\pi\right),$$

we find that

$$\ln N\,\mathbf{P}\{\xi_1 \leqslant r\} = \ln N - \frac{1}{2}\ln\ln N - \frac{1}{2}\ln 4\pi +$$
$$+ \alpha[v - (1+v)\ln(1+v)] + \alpha\Big[-\Delta + \Delta\ln(1+v) +$$
$$+ \Delta\ln\left(1 - \frac{\Delta}{1+v}\right) - (1+v)\ln\left(1 - \frac{\Delta}{1+v}\right)\Big] + o(1),$$

from which it follows that

$$\ln N\mathbf{P}\{\xi_1 \leqslant r\} = -z + o(1).$$

As in the investigation of $\eta_{(N)}$, Lemmas 10, 11, and 12 enable us to prove Theorems 5, 6, and 7 on the limit behavior of the minimal term of an order sequence.

§7. Further Results. References.

Theorem 2.1, the main result of Section 2, is proved in Sevast'yanov and Chistyakov [47]; in the same paper, the generating functions (1.27), and asymptotic expressions for the moments are obtained. The theorem on convergence of the distribution of $\mu_r(n, N)$ to a Poisson distribution in the right-hand r-domain is due to Erdös and Rényi [73]; the proof of convergence to a Poisson distribution in the left-hand r-domain can be found in Kolchin [27]. Asymptotic normality of $\mu_r(n, N)$ in the central domain as well as in the left-hand and the right-hand r-domains is proved by Békéssy [60]. The method of moments used in proving asymptotic normality of $\mu_r(n, N)$ was described in Harris and Park [80]. Local Theorems 3.2, 3.3, and 3.6 are obtained by means of the saddle-point method in Kolchin [27, 28]. The peculiarities of the behavior of the distribution of $\mu_1(n, N)$ as $\alpha \to 0$ is discussed in [28] with the aid of the saddle-point method. In addition to the results given in Section 4, we note that if $\alpha \to 0$ as $n, N \to \infty$, then

$$\sum_{k=0}^{\infty} \mathbf{P}\{\mu_1 = n - 2k\} = \frac{1}{2} + \frac{1}{2} e^{-\frac{1}{3}\alpha^3 N} + o(1),$$

$$\sum_{k=0}^{\infty} \mathbf{P}\{\mu_1 = n - 2k - 1\} = \frac{1}{2} - \frac{1}{2} e^{-\frac{1}{3}\alpha^3 N} + o(1),$$

as was shown in [28].

Thus, as noted in Section 4, if $\alpha^3 N \to \infty$, the distribution of $n-\mu_1$ is concentrated on the lattice of all nonnegative integers. If $\alpha^3 N \to 0$, the distribution of $n-\mu_1$ is asymptotically concentrated on the lattice of nonnegative even integers. Finally, for intermediate values of $\alpha^3 N$ the distribution jumps from one lattice to another. In such cases, the conditional distributions of the random variables $n-\mu_1$ approximate Poisson distributions under the condition that $n-\mu_1$ assumes even or odd values. For more detail, see [28].

The rate of convergence of distributions of $\mu_r(n, N)$ to limit distributions for any fixed $r=0, 1, \ldots$ was studied by Kolchin in [27, 28].

These investigations are similar to those of the problem of the convergence rate of a binomial distribution made by Prokhorov in [41]. The results obtained in his paper are presented in Section 5.

The approach used in Sections 3 and 4 for proving local limit theorems and used in Section 5 for refining these theorems was suggested by Kolchin in [29].

The study of an order series $\eta_{(1)}, \ldots, \eta_{(N)}$ in an equiprobable scheme of allocation of particles was initiated by Viktorova and Sevast'yanov in [11], in which Theorems 6.1 and 6.2 were obtained. Theorem 6.3 is due to Viktorova [9]. The terms of an order series when $\alpha/\ln N$ tends to a constant value was investigated thoroughly by Ivchenko [19]. The investigation of an order series reduces in [9, 11, 19] to the investigation of the asymptotic behavior of random variables $\mu_0(n, N)+\ldots+\mu_r(n, N)$. Section 6 follows Kolchin [30] and uses the representation given by (3.3) of a polynomial distribution in the form of a conditional distribution of independent Poisson random variables with fixed sum. We note that such a representation was used by Dwass in [72] in investigating reverse problems of allocation, by Steck in [97] in investigating a statistic χ^2, by Kolchin in [32] in obtaining a local limit theorem for a statistic $\frac{1}{2}\sum_{i=1}^{N}\eta_i(\eta_i-1)$, where $\eta_1, \ldots, \eta_N$ are occupancies of cells in a nonequiprobable scheme of allocation.

As shown in Section 6, the distribution of the maximum occupancy in an equiprobable scheme of allocation of particles coincides asymptotically with the distribution of the maximum occupancy of N independent random variables $\xi_1, \ldots, \xi_N$ having a Poisson distribution with a parameter α. Kolchin studied in [30] the behavior of an order series in a scheme of series for Poisson random variables $\xi_1, \ldots, \xi_N$ with different modes of variation of the parameter α; the property of unlimited divisibility of a Poisson distribution was used in a significant way.

In order to find the quantiles of a Poisson distribution, the random variable was represented (with integral α) as the sum of independent Poisson random variables with a single parameter, theorems on large deviations being applied to this sum.

The generalization of such an approach to the investigation of an order sequence in a scheme of series can be found in G. I. Ivchenko [21], where the behavior of an order sequence constructed for independent uniformly distributed random variables $\xi_1, \ldots, \xi_N$ was examined in the case when $\xi_1 = X_1 + \ldots + X_n$, where $X_1, \ldots, X_n$ are independent and identically distributed with a distribution function $F(x)$ independent of the parameters n and N. Particular cases of this scheme, where ξ_1 has a Poisson distribution and a gamma-distribution with parameters independent of N are examined by Ivchenko in [22]. In [23], Ivchenko studied (in an equiprobable scheme of allocation of particles) the behavior of an order sequence made of occupancies of cells at some random instant of time.

Chapter III

MULTINOMIAL ALLOCATIONS

§1. Generating Functions and Moments

In this chapter, we shall assume that each of n particles gets into the ith cell with probability a_i, $i=1, 2, \ldots, N$ independently of the remaining particles. Let us denote by $\mu_r(n, a_1, a_2, \ldots, a_N)$ (or simply $\mu_r(n)$, μ_r) the number of cells containing exactly r particles. We introduce the generating functions

$$\Phi_{r_1 \ldots r_s}(z, x, a_1, \ldots, a_N) =$$
$$= \sum_{n=0}^{\infty} \frac{(zN)^n}{n!} \sum_{k_1, \ldots, k_s=0}^{\infty} x^k \mathbf{P}\{\mu_{r_i}(n, a_1, \ldots, a_N) = k_i,\ i=1, \ldots, s\}, \tag{1}$$

where $x = (x_1, \ldots, x_s)$ and $x^k = x_1^{k_1} x_2^{k_2} \ldots x_s^{k_s}$.

Theorem 1. *The generating function given by* (1) *is of the following form:*

$$\Phi_{r_1 \ldots r_s}(z, x, a_1, \ldots, a_N) = \\ = \prod_{k=1}^{N} \left[e^{zNa_k} + \sum_{i=1}^{s} \frac{(zNa_k)^{r_i}}{r_i\,!} (x_i - 1) \right]. \quad (2)$$

PROOF. We divide the N cells into two groups: group one consisting of cells numbered $1, 2, \ldots, N_1$, and group two consisting of the remaining cells. The probability that n_1 particles fall into the group with N_1 cells when n particles are thrown is equal to $C_n^{n_1} p^{n_1} q^{n_2}$, where

$$p = a_1 + \ldots + a_{N_1}, \quad q = 1 - p, \quad n_2 = n - n_1.$$

If the numbers of particles in group one and in group two are fixed, then the particles are distributed independently of each other with probabilities $a_1/p, \ldots, a_{N_1}/p$ for group one and probabilities $a_{N_1+1}/q, \ldots, a_N/q$ for group two. By the law of composition of probabilities, we have

$$\mathbf{P}\{\mu_{r_i}(n, a_1, \ldots, a_N) = k_i, \ i = 1, \ldots, s\} = \\ = \sum_{n_1+n_2=n} \sum_{\substack{k_{i1}+k_{i2}=k_i \\ i=1,\ldots,s}} C_n^{n_1} p^{n_1} q^{n_2} \times \\ \times \mathbf{P}\left\{\mu_{r_i}\left(n_1, \frac{a_1}{p}, \ldots, \frac{a_{N_1}}{p}\right) = k_{i1}, \ i = 1, \ldots, s\right\} \times \\ \times \mathbf{P}\left\{\mu_{r_i}\left(n_2, \frac{a_{N_1+1}}{q}, \ldots, \frac{a_N}{q}\right) = k_{i2}, \ i = 1, \ldots, s\right\}.$$

Multiplying both sides of this equality by $N^n z^n x^k / n!$ and summing over $n, k_1, \ldots, k_s$, we obtain

$$\Phi_{r_1 \ldots r_s}(z, x, a_1, \ldots, a_N) = \\ = \Phi_{r_1 \ldots r_s}\left(p \frac{N}{N_1} z, x, \frac{a_1}{p}, \ldots, \frac{a_{N_1}}{p}\right) \Phi_{r_1 \ldots r_s}\left(q \frac{N}{N_2} z, x, \frac{a_{N_1+1}}{q}, \ldots, \frac{a_N}{q}\right). \quad (3)$$

For $N_1 = 1$, we can write (3) as

$$\Phi_{r_1 \ldots r_s}(z, x, a_1, \ldots, a_N) =$$
$$= \Phi_{r_1 \ldots r_s}(a_1 zN, x, 1)\, \Phi_{r_1 \ldots r_s}\left((1 - a_1)\frac{N}{N-1} z, x, \frac{a_2}{1-a_1}, \ldots, \frac{a_N}{1-a_1}\right). \tag{4}$$

Applying (4) (replacing N with $N-1$ and z with $\frac{(1-a_1)\,Nz}{N-1}$) to the second factor in (4), we get

$$\Phi_{r_1 \ldots r_s}\left((1 - a_1)\frac{N}{N-1} z, x, \frac{a_2}{1-a_1}, \ldots, \frac{a_N}{1-a_1}\right) =$$
$$= \Phi_{r_1 \ldots r_s}(a_2 zN, x, 1) \times$$
$$\times \Phi_{r_1 \ldots r_s}\left(\frac{(1 - a_1 - a_2)\, zN}{N-2}, x, \frac{a_3}{1-a_1-a_2}, \ldots, \frac{a_N}{1-a_1-a_2}\right).$$

Continuing the process of singling out the factors corresponding to a single cell, we find from the chain of equations obtained that

$$\Phi_{r_1 \ldots r_s}(z, x, a_1, \ldots a_N) = \prod_{k=1}^{N} \Phi_{r_1 \ldots r_s}(a_k Nz, x, 1). \tag{5}$$

Since

$$\mathbf{P}\{\mu_{r_i}(n, 1) = 0,\ i = 1, \ldots, s\} = 1, \quad n \neq r_i,\ i = 1, \ldots, s,$$

and

$$\mathbf{P}\{\mu_{r_i}(n, 1) = 1,\ \mu_{r_j}(n, 1) = 0,\ j \neq i\} = 1, \quad n = r_i,$$

we have

$$\Phi_{r_1 \ldots r_s}(a_k Nz, x, 1) = e^{a_k Nz} + \sum_{i=1}^{s} \frac{(a_k Nz)^{r_i}}{r_i!}(x_i - 1).$$

The assertion of the theorem follows from this and (5).

We shall need also the formulas for the generating functions of the variables $\mu_{r_1}, \ldots, \mu_{r_s}$.

Theorem 2. *The generating function*

$$F_{n,r_1\ldots r_s}(x, a_1, \ldots, a_N) = \mathbf{M}\prod_{j=1}^{s} x_j^{\mu_{r_j}(n,a_1,\ldots,a_N)}$$

is given by the formula

$$F_{n,r_1\ldots r_s}(x, a_1, \ldots, a_N) =$$

$$= \frac{n!}{N^n 2\pi i}\oint \prod_{k=1}^{N}\left(e^{a_k N z} + \sum_{l=1}^{s}\frac{(a_k N z)^{r_l}}{r_l!}(x_l - 1)\right)\frac{dz}{z^{n+1}}, \tag{6}$$

where the integral is over any closed contour encircling the point $z=0$.

PROOF. We can write (1) as

$$\Phi_{r_1\ldots r_s}(z, x, a_1, \ldots, a_N) = \sum_{n=0}^{\infty}\frac{N^n z^n}{n!}F_{n,r_1\ldots r_s}(x, a_1, \ldots, a_N).$$

The assertion of the theorem follows from this and (2).

The formula

$$\mathbf{M}x^{\mu_0(n)} = \frac{n!}{N^n 2\pi i}\oint\prod_{k=1}^{N}(e^{a_k N z} + x - 1)\frac{dz}{z^{n+1}} \tag{7}$$

is a particular case of (6).

Using (2), we obtain, as in Section 1, Ch. 2, formulas analogous to (2.1.3)-(2.1.5):

$$\mathbf{M}\mu_r = \sum_{k=1}^{N} C_n^r a_k^r (1 - a_k)^{n-r}, \tag{8}$$

$$\mathbf{M}\mu_r^{[2]} = \sum_{k \neq l} \frac{n!}{(r!)^2 (n-2r)!} a_k^r a_l^r (1 - a_k - u_l)^{n-2r}, \quad (9)$$

$$\mathbf{M}\mu_r\mu_t = \sum_{k \neq l} \frac{n!}{r!t!(n-r-t)!} a_k^r a_l^t (1 - a_k - a_l)^{n-r-t}, r \neq t. \quad (10)$$

Similarly, we can calculate the factorial moment μ_r:

$$\mathbf{M}\mu_r^{[k]} = \frac{n^{[kr]}}{(r!)^k} \sum_{l_1 \ldots l_k}^{*} a_{l_1}^r \ldots a_{l_k}^r \left(1 - a_{l_1} - \ldots - a_{l_k}\right)^{n-kr}. \quad (11)$$

Here and below, we denote by $\sum^*$ the sum over all $N^{[k]}$ k-tuples $(l_1, l_2, \ldots, l_k)$ in which all the l_i are different.

In this chapter, we shall discuss the asymptotic behavior of various characteristics of the variables μ_r as n, $N \to \infty$. If the probabilities $a_1, a_2, \ldots, a_N$ change in an arbitrary manner as N increases, there will of course be a great many types of asymptotic behavior of the distribution of μ_r. By imposing some natural restrictions on the a_i, however, we can preserve the essential characteristics of the central domain as well as the right-hand and left-hand r-domains.

Parameters n, $N \to \infty$ and a_k are said to *vary in the central domain* if there exists positive constants C and $\alpha_0 < \alpha_1$ such that

$$N a_k \leqslant C, \quad \alpha_0 \leqslant \alpha = \frac{n}{N} \leqslant \alpha_1, \quad n, N \to \infty. \quad (12)$$

Let us define

$$\underline{a} = \min_{1 \leqslant k \leqslant N} a_k, \quad \bar{a} = \max_{1 \leqslant k \leqslant N} a_k.$$

We shall say that n, $N \to \infty$ and a_k *vary in the left-hand r-domain* (for $r \geqslant 2$) if

$$n\bar{a} \to 0 \quad \text{and} \quad \mathbf{M}\mu_r \to \lambda < \infty. \quad (13)$$

Just as in an equiprobable scheme, we identify the left-hand r-domains (for $r=0$ and $r=1$) with the left-hand 2-domain. The *right-hand r domain* is defined for any $r=0, 1, 2, \ldots$ by the relations

$$na \to \infty, \quad \mathbf{M}\mu_r \to \lambda < \infty. \tag{14}$$

Let us investigate first the behavior of $\mathbf{M}\mu_0$, $\mathbf{D}\mu_0$ in the central domain. From (8) and (9) we find (with $r=0$) that

$$\mathbf{M}\mu_0 = \sum_{i=1}^{N} (1-a_i)^n,$$

$$\mathbf{D}\mu_0 = \sum_{i=1}^{N} (1-a_i)^n - \\ - \sum_{i,j=1}^{N} (1-a_i)^n (1-a_j)^n + \sum_{i \neq j} (1-a_i-a_j)^n, \tag{15}$$

or

$$\mathbf{D}\mu_0 = \sum_{i=1}^{N} (1-a_i)^n - \sum_{i=1}^{N} (1-2a_i)^n - \\ - \sum_{i,j=1}^{N} [(1-a_i-a_j+a_ia_j)^n - (1-a_i-a_j)^n]. \tag{16}$$

Theorem 3. *In the central domain* (12), *we have the asymptotic formulas*

$$\mathbf{M}\mu_0 = Nm_N + O(1), \quad \mathbf{D}\mu_0 = N\sigma_N^2 + O(1),$$

where

$$m_N = \frac{1}{N} \sum_{k=1}^{N} e^{-\alpha N a_k}, \tag{17}$$

$$\sigma_N^2 = \frac{1}{N}\sum_{k=1}^{N} e^{-\alpha N a_k} - \frac{1}{N}\sum_{k=1}^{N} e^{-2\alpha N a} - $$
$$- \alpha\left(\frac{1}{N}\sum_{k=1}^{N} N a_k e^{-\alpha N a_k}\right)^2. \qquad (18)$$

PROOF. Under the conditions of (12), we have

$$(1 - a_i)^n = e^{\alpha N \ln(1-a_i)} = e^{-\alpha N a_i} + O\left(\frac{1}{N}\right),$$
$$(1 - 2a_i)^n = e^{-2\alpha N a_i} + O\left(\frac{1}{N}\right),$$
$$(1 - a_i - a_j + a_i a_j)^n =$$
$$= e^{-\alpha N a_i} \cdot e^{-\alpha N a_j}\left[1 + \alpha N a_i a_j - \frac{\alpha N}{2}(a_i + a_j)^2\right] + O\left(\frac{1}{N^2}\right),$$
$$(1 - a_i - a_j)^n =$$
$$= e^{-\alpha N a_i} e^{-\alpha N a_j}\left[1 - \frac{\alpha N}{2}(a_i + a_j)^2\right] + O\left(\frac{1}{N^2}\right).$$

From these formulas and formulas (15) and (16) the assertion of the theorem follows.

Theorem 4. *In the central domain* (12), *the variable* σ_N^2 *defined by* (18) *satisfies the inequality*

$$\sigma_N^2 \geqslant \varepsilon > 0,$$

where ε *is a constant.*

PROOF. Applying in (18) the Cauchy-Bunyakovskiy inequality

$$\left(\sum_{k=1}^{N} e^{-\alpha N a_k} a_k\right)^2 \leqslant \left(\sum_{k=1}^{N} a_k\right)\left(\sum_{k=1}^{N} e^{-2\alpha N a_k} a_k\right),$$

we obtain

$$\sigma_N^2 \geqslant \frac{1}{N} \sum_{k=1}^{N} \sigma(\alpha N a_k), \tag{19}$$

where $\sigma(\gamma) = e^{-\gamma}[1-(1+\gamma)e^{-\gamma}]$.

Let N_δ be the number of values of k for which

$$a_k \geqslant \frac{\delta}{N}, \quad 0 < \delta < 1, \tag{20}$$

and let us define

$$\sigma_0 = \min_{0<\alpha\delta\leqslant\gamma\leqslant\alpha C} \sigma(\gamma),$$

where C is the constant referred to in (12). Obviously, $\sigma_0 > 0$. Since

$$1 = \sum_{k=1}^{N} a_k \leqslant \frac{\delta}{N}(N - N_\delta) + \frac{C}{N} N_\delta,$$

we have

$$\frac{N_\delta}{N} \geqslant \frac{1-\delta}{C-\delta}. \tag{21}$$

Summing in (19) over those k that satisfy (20) and replacing $\sigma(\alpha N a_k)$ by σ_0, we have

$$\sigma_N^2 \geqslant \frac{N_\delta}{N} \sigma_0 \geqslant \frac{1-\delta}{C-\delta} \sigma_0 > 0.$$

Theorem 5. *In the central domain* (12), *we have the asymptotic formulas*

$$\mathbf{M}\mu_r = N\widetilde{m}_r + O(1), \quad \mathbf{Cov}(\mu_r, \mu_t) = N\widetilde{\sigma}_{rt} + O(1),$$

where

$$\widetilde{m}_r = \frac{1}{N}\sum_{k=1}^{N} p_{r,k},$$

$$\widetilde{\sigma}_{rr} = \frac{1}{N}\sum_{k=1}^{N} p_{r,k} - \frac{1}{N}\sum_{k=1}^{N} p_{r,k}^2 - \frac{1}{\alpha}\left[\frac{1}{N}\sum_{k=1}^{N} p_{r,k}(\alpha N a_k - r)\right]^2,$$

$$\widetilde{\sigma}_{rt} = -\frac{1}{N}\sum_{k=1}^{N} p_{r,k}p_{t,k} -$$

$$-\frac{1}{\alpha}\left(\frac{1}{N}\sum_{k=1}^{N} p_{r,k}(\alpha N a_k - r)\right)\left(\frac{1}{N}\sum_{k=1}^{N} p_{t,k}(\alpha N a_k - t)\right),$$

$$p_{r,k} = \frac{(\alpha N a_k)^r}{r!} e^{-\alpha N a_k}. \tag{22}$$

This theorem is proved with the aid of (8)-(10) in the same way as Theorem 3.

Sometimes we shall impose on the a_k conditions stronger than (12), namely,

$$\alpha_0 \leqslant \frac{n}{N} = \alpha \leqslant \alpha_1, \quad a_k = \frac{1}{N}\left(1 + \frac{b_k^{(N)}}{\sqrt[4]{N}}\right),$$

$$k = 1, \ldots, N, \quad |b_k^{(N)}| \leqslant K < \infty,$$

$$b_N^2 = \frac{1}{N}\sum_{k=1}^{N} (b_k^{(N)})^2 \to b^2 > 0 \text{ as } n, N \to \infty. \tag{23}$$

Theorem 6. *If conditions* (23) *are satisfied, then, as* $n, N \to \infty$,

$$\widetilde{m}_r = p_r\left(1 + \frac{b_N^2}{2\sqrt{N}}[(\alpha - r)^2 - r]\right) + O\left(\frac{1}{N^{3/4}}\right), \tag{24}$$

$$\widetilde{\sigma}_{rt} = \sigma_{rt} + O\left(\frac{1}{\sqrt{N}}\right),$$

where $p_r = \frac{\alpha^r}{r!}e^{-\alpha}$ *and the variables* σ_{rt} *and* b_N^2 *are defined by* (2.1.9) *and* (23).

PROOF. Substituting a_k from (23) into (22), we obtain

$$p_{r,k} = \frac{\alpha^r}{r!} e^{-\alpha}\left(1 + \frac{b_k^{(N)}}{\sqrt[4]{N}}\right)^r \exp\left(-\frac{\alpha b_k^{(N)}}{\sqrt[4]{N}}\right) =$$

$$= p_r\left[1 - \frac{(\alpha - r)}{\sqrt[4]{N}} b_k^{(N)} + \frac{\left(b_k^{(N)}\right)^2}{2\sqrt{N}}[(\alpha - r)^2 - r]\right] + O\left(\frac{1}{N^{3/4}}\right). \tag{25}$$

Since $\sum_{k=1}^{N} a_k = 1$, it follows from (23) that

$$\sum_{k=1}^{N} b_k^{(N)} = 0. \tag{26}$$

The assertion of the theorem follows from formulas (25) and (26) and Theorem 4.

We consider now the left-hand r-domain for $r \geqslant 2$. Since $\bar{a} \geqslant 1/N$, it follows from (13) that $\alpha = n/N \leqslant n\bar{a} \to 0$. Further, from the inequalities $(1-\bar{a})^{n-r} \leqslant (1-a_i)^{n-r} \leqslant 1$ and from the relation $\lim_{n\to\infty} (1-\bar{a})^{n-r} = 1$, we have

$$\mathbf{M}\mu_r \sim \frac{n^r}{r!} \sum_{i=1}^{N} a_i^r \to \lambda. \tag{27}$$

Let $\bar{\mu}_r = \sum_{k>r} \mu_r$. From the obvious inequalities

$$\mathbf{M}\bar{\mu}_r = \sum_{k=r+1}^{n} \mathbf{M}\mu_k = \sum_{k=r+1}^{n} \sum_{i=1}^{N} C_n^k a_i^k (1 - a_i)^{n-k} \leqslant$$

$$\leqslant \sum_{k=r+1}^{\infty} (n\bar{a})^{k-r} \frac{n^r}{r!} \sum_{i=1}^{N} a_i^r = \frac{n\bar{a}}{1 - n\bar{a}} (\lambda + o(1))$$

we see that $\mathbf{M}\bar{\mu}_r \to 0$ and then, by Chebyshev's inequality,

$$\mathbf{P}\{\bar{\mu}_r > 0\} \leqslant \mathbf{M}\bar{\mu}_r \to 0, \tag{28}$$

i.e., in the left-hand r-domain, all cells are, with probability approaching 1, occupied by no more than r particles. Hence, in particular, in the left-hand 2-domain, with probability approaching 1, the equations $\mu_0+\mu_1+\mu_2=N$ and $\mu_1+2\mu_2=n$ are satisfied. These can be expressed in the equivalent forms

$$\mu_0=N-n+\mu_2, \tag{29}$$

$$\mu_1=n-2\mu_2. \tag{30}$$

In the right-hand r-domain, by virtue of (14) and the fact that $a\leqslant 1/N$, we have $\alpha=n/N\geqslant n\underline{a}\to\infty$. Let $\underline{\mu}_r=\sum\limits_{k<r}\mu_k$. From the inequalities

$$\mathbf{M}\underline{\mu}_r=\sum_{k=0}^{r-1}\mathbf{M}\mu_k=\sum_{k=0}^{r-1}\sum_{i=1}^{N}C_n^k a_i^k(1-a_i)^{n-k}\leqslant$$

$$\leqslant\sum_{k=0}^{r-1}\sum_{i=1}^{N}\frac{C_n^k}{C_n^r}\,\frac{C_n^r a_i^r(1-a_i)^{n-r}}{\underline{a}^{r-k}}=O\left(\frac{1}{n\underline{a}}\right)\to 0$$

we see that $\mathbf{M}\underline{\mu}_r\to 0$, and, again by Chebyshev's inequality,

$$\mathbf{P}\{\underline{\mu}_r>0\}\leqslant\mathbf{M}\underline{\mu}_r\to 0, \tag{31}$$

i.e., in the right-hand r-domain, with probability tending to 1, each cell contains no fewer than r particles.

§2. Poisson Limit Theorem for the Sums of Indicators

This section is devoted to some auxiliary material. In proving limit theorems, it is sometimes convenient to represent a random variable ξ as the sum of indicators, i.e., as

$$\xi=\eta_1+\eta_2+\ldots+\eta_n, \tag{1}$$

where the η_i assume the values 1 and 0 only and are, generally speaking, mutually dependent. We formulate here some general conditions under which the limit distribution of ξ as $n\to\infty$ is a Poisson distribution.

We denote in (1)

$$b_{i_1 i_2 \ldots i_r} = \mathbf{P}\{\eta_{i_1} = \eta_{i_2} = \ldots = \eta_{i_r} = 1\},$$

where no two indices $1 \leqslant i_k \leqslant n$ are equal and where $r = 1, 2, \ldots, n$. Further, for any $r = 2, 3, \ldots$ and $n = 2, 3, \ldots$, we introduce "exclusive" sets $I_r(n)$ of order r-tuples $(i_1, i_2, \ldots, i_r)$, where no two indices $1 \leqslant i_k \leqslant n$ are equal.

Theorem 1. *Let a random variable* ξ *be represented by* (1). *If the probabilities* $b_{i_1 \ldots i_r}$ *satisfy the conditions*

$$\max_{1\leqslant i\leqslant n} b_i \to 0, \qquad \lim_{n\to\infty} \sum_{i=1}^{n} b_i = \lambda, \tag{2}$$

for any $r = 2, 3, \ldots$ *and* $n = 2, 3, \ldots$ *and if there exist exclusive sets* $I_r(n)$ *such that*

$$\lim_{n\to\infty} \sum_{(i_1 \ldots i_r)\in I_r(n)} b_{i_1 \ldots i_r} = \lim_{n\to\infty} \sum_{(i_1 \ldots i_r)\in I_r(n)} b_{i_1} b_{i_2} \ldots b_{i_r} = 0 \tag{3}$$

and

$$\lim_{n\to\infty} \max_{(i_1,\ldots,i_r)\in I_r(n)} \left| \frac{b_{i_1 i_2 \ldots i_r}}{b_{i_1} b_{i_2} \ldots b_{i_r}} - 1 \right| = 0, \tag{4}$$

then

$$\lim_{n\to\infty} \mathbf{P}\{\xi = k\} = \frac{\lambda^k}{k!} e^{-\lambda}, \qquad k = 0, 1, 2, \ldots . \tag{5}$$

PROOF. Let us apply the method of moments. We shall have proved (5) if we can establish that

$$\lim_{n\to\infty} \mathbf{M}\xi^{[r]} = \lambda^r, \quad r = 1, 2, \ldots \tag{6}$$

It is easy to verify that

$$\mathbf{M}\xi^{[r]} = \sum_{i_1 \ldots i_r}^{*} b_{i_1 \ldots i_r}, \tag{7}$$

where $\sum^*$ is taken over all $n^{[r]}$ ordered sets of pairwise distinct indices $i_1, i_2, \ldots, i_r$. We break the summation (7) into two summations:

$$\sum_{i_1 \ldots i_r}^{*} b_{i_1 \ldots i_r} = \sum_{(i_1 \ldots i_r) \notin I_r(n)} b_{i_1 \ldots i_r} + \sum_{(i_1 \ldots i_r) \in I_r(n)} b_{i_1 \ldots i_r}. \tag{8}$$

It follows from (4) that for any $\varepsilon > 0$ there exists an n_0 such that, for every $n > n_0$ and for every $(i_1 \ldots i_r) \notin I_r(n)$,

$$b_{i_1} \ldots b_{i_r}(1 - \varepsilon) \leqslant b_{i_1 \ldots i_r} \leqslant b_{i_1} \ldots b_{i_r}(1 + \varepsilon). \tag{9}$$

Using (9), we obtain

$$(1 - \varepsilon) \sum_{(i_1 \ldots i_r) \notin I_r(n)} b_{i_1} \ldots b_{i_r} \leqslant \sum_{(i_1 \ldots i_r) \notin I_r(n)} b_{i_1 \ldots i_r} \leqslant$$
$$\leqslant (1 + \varepsilon) \sum_{(i_1 \ldots i_r) \notin I_r(n)} b_{i_1} \ldots b_{i_r}. \tag{10}$$

By virtue of (3), the second summation in (8) approaches zero as $n \to \infty$. Furthermore,

$$\sum_{(i_1 \ldots i_r) \notin I_r(n)} b_{i_1} \ldots b_{i_r} =$$
$$= \left(\sum_{i=1}^{n} b_i \right)^r - \sum_{(i_1 \ldots i_r) \in I_r(n)} b_{i_1} \ldots b_{i_r} - \sum_{i_1 \ldots i_r}' b_{i_1} \ldots b_{i_r}, \tag{11}$$

where the last summation $\sum'$ is over all r-tuples $(i_1, i_2, \ldots, i_r)$ that have at least two equal components. It can easily be seen that

$$\left|\sum_{i_1 \ldots i_r}' b_{i_1} \ldots b_{i_r}\right| \leqslant \max_{1 \leqslant i \leqslant r} b_i \left[\sum_{i=1}^{n} b_i\right]^{r-1} \to 0. \tag{12}$$

From (2), (8), (10), (11), and (12) we deduce (6); i.e., the factorial moments of ξ converge to the factorial moments λ^r of the Poisson distribution, proving the theorem. We shall use this result in the next section.

§3. Limit Distributions in the Left- and Right-hand Domains

Let us prove the Poisson limit theorems for μ_r.

Theorem 1. *In the left-hand r-domain, that is, under the conditions* (1.13), *we have for, any* $r \geqslant 2$,

$$\lim \mathbf{P}\{\mu_r = k\} = \frac{\lambda^k}{k!} e^{-\lambda}, \qquad k = 0, 1, \ldots, \tag{1}$$

where, by (1.27),

$$\lim_{n\to\infty} \frac{n^r}{r!} \sum_{i=1}^{N} a_i^r = \lambda. \tag{2}$$

Proof. Let us apply the method of moments. From (1.13) and (1.27) we see that $\mathbf{M}\mu_r \to \lambda$. From (1.11) we see that

$$\mathbf{M}\mu_r^{[k]} \leqslant \frac{n^{kr}}{(r!)^k} \left(\sum_{i=1}^{N} a_i^r\right)^k \to \lambda^k. \tag{3}$$

Using in (1.11) the estimate from below

$$(1-k\bar{a})^n \leqslant (1-a_{l_1}-\ldots-a_{l_k})^{n-kr},$$

we get

$$(1-k\bar{a})^n \frac{n^{[kr]}}{(r!)^k}\left[\left(\sum_{l=1}^{N} a_l^r\right)^k - \sum_{l_1..l_k}{}' a_{l_1}^r \cdots a_{l_k}^r\right] \leqslant \mathbf{M}\mu_r^{[k]}, \tag{4}$$

where Σ' denotes summation over all combinations of indices $l_1, l_2, \ldots, l_k$ among which there is at least one pair of equal indices. Since

$$\lim_{n\to\infty}(1-k\bar{a})^n = 1 \quad \text{and} \quad \sum_{l_1\ldots l_k}{}' a_{l_1} \ldots a_{l_k} \leqslant \bar{a}^r\left(\sum_{l=1}^{N} a_i^r\right)^{k-1},$$

it follows from (4) that

$$\lambda^k \leqslant \lim \mathbf{M}\mu_r^{[k]}. \tag{5}$$

From (3) and (5), we get $\lim \mathbf{M}\mu_r^{[k]} = \lambda^k$, thus proving the theorem.

Theorem 2. *In the left-hand r-domain, for* $r \leqslant 2$, *if conditions* (1.13) *are satisfied, the random variables*

$$\mu_0 - (N-n), \quad \frac{n-\mu_1}{2}, \quad \mu_2$$

all have the same limiting Poisson distribution with parameter

$$\lambda = \lim \frac{n^2}{2} \sum_{i=1}^{N} a_i^2.$$

The proof for $r=2$ is given in Theorem 1. The remaining assertions of the theorem follow from the fact that equations (1.29) and (1.30) are satisfied in the left-hand 2-domain with probability approaching 1.

To prove the limit theorem in the right-hand domain, we shall make use of Theorem 2.1.

Theorem 3. *In the right-hand r-domain, that is under conditions* (1.14), *for any r,*

$$\lim \mathbf{P}\{\mu_r = k\} = \frac{\lambda^k}{k!} e^{-\lambda}, \qquad k = 0, 1, 2, \ldots, \tag{6}$$

where

$$\lambda = \lim \mathbf{M}\mu_r. \tag{7}$$

PROOF. Let us represent μ_r as the sum of indicators

$$\mu_r = \eta_1 + \eta_2 + \ldots + \eta_N, \tag{8}$$

where η_i is equal to 1 if there are r particles in the ith box and equal to zero otherwise. The probabilities $b_{i_1 \ldots i_k} = \mathbf{P}\{\eta_{i_1} = \cdots = \eta_{i_k} = 1\}$ are defined by the formulas

$$b_{i_1 \ldots i_k} = \frac{n^{[kr]}}{(r!)^k} a_{i_1}^r \ldots a_{i_k}^r \left(1 - a_{i_1} - \ldots - a_{i_k}\right)^{n-kr}. \tag{9}$$

As proved in Section 1, $\alpha = n/N \to \infty$ in the right-hand r-domain, i.e., $N = o(n)$. Let $I_1(n)$ denote a subset of $\{1, 2, \ldots, N\}$ such that $i \in I_1(n)$ if an only if $a_i > n^{-3/4}$. We shall include in the exclusive set $I_k(n)$ those k-tuples $(i_1, \ldots, i_k)$ for which at least one $i_k \in I_1(n)$ and only those. We show now that the probabilities given by (9) satisfy the conditions of Theorem 2.1. First we note that since, for $x > r/n$, the function $x^r(1-x)^{n-r}$ is decreasing, it follows that, for $r < \sqrt[4]{n}$, the inequalities $r/n < n^{-3/4} < a_i$ lead to the estimate

$$\sum_{i \in I_1(n)} b_i \leqslant \frac{n^r}{r!} \sum_{i \in I_1(n)} n^{-3r/4} (1 - n^{-3/4})^{n-r} \leqslant$$
$$\leqslant N \frac{n^{r/4}}{r!} e^{-n^{1/4} + rn^{-3/4}} = o\left(n^{1+\frac{r}{4}} e^{-n^{1/4}}\right) \to 0.$$

From this, we see that, for any $r=2, 3, \ldots,$

$$\sum_{i_1\ldots i_k \in I_k(r)} b_{i_1} \ldots b_{i_k} \leqslant k\left(\sum_{i=1}^{N} b_i\right)^{k-1} \sum_{i \in I_1(n)} b_i \to 0 \qquad (10)$$

as $n,\ N \to \infty$. Since

$$b_{i_1 \ldots i_k} \leqslant n^{kr}\left(1 - a_{i_1} - \ldots - a_{i_k}\right)^{n-r}$$

and $a_{i_1} + \ldots + a_{i_k} > n^{-3/4}$ for $(i_1, \ldots, i_k) \in I_k(n)$, it follows that

$$\sum_{(i_1 \ldots i_k) \in I_k(n)} b_{i_1 \ldots i_k} \leqslant n^{kr}(1 - n^{-3/4})^{n-r} \leqslant$$
$$\leqslant n^{kr} e^{-n^{1/4} + rn^{-3/4}} \to 0$$

as $n \to \infty$.

This relation together with (10) shows that condition (2.3) of Theorem 2.1 is satisfied. From now on, we shall use the fact that for $0 \leqslant x \leqslant 1/2$

$$-x - x^2 \leqslant \ln(1-x) \leqslant -x. \qquad (11)$$

Since $a_{i_1} + \ldots + a_{i_k} \leqslant kn^{-3/4}$ for $(i_1, \ldots, i_k) \notin I_k(n)$, we see that $a_{i_1} + \ldots + a_{i_k} \leqslant 1/2$ for $n \geqslant (2k)^{4/3}$ and therefore we can apply (11).

Hence, for $(i_1 \ldots i_k) \in I_k(n)$ and $n \geqslant (2k)^{4/3}$, we have

$$\frac{n^{[kr]}}{(r!)^k} a_{i_1}^r \ldots a_{i_k}^r e^{-(n-kr)[(a_{i_1}+\ldots+a_{i_k}) + (a_{i_1}+\ldots+a_{i_k})^2]} \leqslant b_{i_1 \ldots i_k} \leqslant$$
$$\leqslant \frac{n^{[kr]}}{(r!)^k} a_{i_1}^r \ldots a_{i_k}^r e^{-(n-kr)(a_{i_1}+\ldots+a_{i_k})}. \qquad (12)$$

Since $a_{i_l} \leqslant n^{-3/4}$ the left-hand inequality can also be expressed:

$$\frac{n^{[kr]}}{(r!)^k} a_{i_1}^r \ldots a_{i_k}^r e^{-n(a_{i_1}+\ldots+a_{i_r})} e^{-n^{-1/2}k^2} \leqslant b_{i_1 \ldots i_k}. \qquad (13)$$

For $k=1$, (12) and (13) yield the inequalities

$$\frac{n^{[r]}}{r!} a_i^r e^{-na_i-n^{-1/2}} \leqslant b_i \leqslant \frac{n^{[r]}}{r!} a_i^r e^{-(n-r)a_i}. \quad (14)$$

From (12), (13), and (14) we get

$$\frac{n^{[kr]}}{(n^{[r]})^k} e^{-krn^{-3/4}-k^2n^{-1/2}} \leqslant \frac{b_{i_1\ldots i_r}}{b_{i_1}\ldots b_{i_r}} \leqslant$$

$$\leqslant \frac{n^{[kr]}}{(n^{[r]})^k} e^{kn^{-1/2}+k^2rn^{-3/4}},$$

from which condition (2.4) of Theorem 2.1 follows.

The first condition in (2.2) follows from (7) and the fact that

$$b_i = C_n^r a_i^r (1-a_i)^{n-r} \leqslant \frac{(na_i)^r}{r!} e^{-a_i(n-r)} \leqslant e^r \frac{(n\underline{a})^r}{r!} e^{-n\underline{a}} \to 0.$$

We have thus shown that all the conditions of Theorem 2.1 are satisfied, proving (6), and hence the theorem.

§4. A Normal Limit Distribution of the Number of Empty Cells

Let us prove the asymptotic normality of μ_0 in the central domain (1.12). We shall obtain both an integral and a local theorem. We prove first convergence to $e^{-t^2/2}$ (as $n, N \to \infty$) of the characteristic functions

$$\Psi_N(t) = \mathbf{M} \exp\left\{it \frac{\mu_0 - Nm_N}{\sigma_N \sqrt{N}}\right\}, \quad (1)$$

where m_N and σ_N are defined by (1.17) and (1.18).

Theorem 1. *In the central domain defined by* (1.12) *the* *characteristic function* $\Psi_N(t)$ *defined by* (1) *converges to* $e^{-t^2/2}$ *uniformly with respect to* $|t| \leqslant \sqrt[8]{N}$ *as* $n, N \to \infty$.

PROOF. Making use of (1.7), we obtain

$$\Psi_N(t) = e^{-\frac{itm_N\sqrt{N}}{\sigma_N}} \frac{n!}{N^n 2\pi i} \oint \prod_{k=1}^{N} \left(e^{zNa_k} + e^{\frac{it}{\sigma_N\sqrt{N}}} - 1\right) \frac{dz}{z^{n+1}}.$$

We integrate over the contour $|z| = \alpha$. Setting $z = \alpha \exp\left(\frac{iv}{\sqrt{\alpha N}}\right)$ and using Stirling's formula for $n!$, we see that, as $n, N \to \infty$,

$$\Psi_N(t) = \frac{1 + o(1)}{\sqrt{2\pi}} e^{-\frac{itm_N\sqrt{N}}{\sigma_N}} \int_{-\pi\sqrt{\alpha N}}^{\pi\sqrt{\alpha N}} e^{-iv\sqrt{\alpha N}} \prod_{k=1}^{N} A_k(v)\, dv, \quad (2)$$

where

$$A_k(v) = \left(e^{\frac{it}{\sigma_N\sqrt{N}}} - 1\right) e^{-\alpha} + e^{-\alpha} \exp\left(\alpha N a_k e^{\frac{iv}{\sqrt{\alpha N}}}\right). \quad (3)$$

The product in (2) can be expressed as

$$\prod_{k=1}^{N} A_k(v) = \left(\prod_{k=1}^{N} A_{1k}(v)\right) \cdot \left(\prod_{k=1}^{N} A_{2k}(v)\right), \quad (4)$$

where

$$A_{1k}(v) = \exp\left\{\alpha N a_k \left(e^{\frac{iv}{\sqrt{\alpha N}}} - 1\right)\right\}, \quad (5)$$

$$A_{2k}(v) = 1 + \left(e^{\frac{it}{\sigma_N\sqrt{N}}} - 1\right) \exp\left(-\alpha N a_k e^{\frac{iv}{\sqrt{\alpha N}}}\right). \quad (6)$$

To verify (4), we take the second term outside the parentheses in (3) and note that

$$\prod_{k=1}^{N} \exp\left\{\alpha\left(Na_k e^{\frac{iv}{\sqrt{\alpha N}}} - 1\right)\right\} = \prod_{k=1}^{N} \exp\left\{\alpha N a_k \left(e^{\frac{iv}{\sqrt{\alpha N}}} - 1\right)\right\}.$$

Let us partition the integration domain in (2) into three domains

$$S_1 = \{v : \delta\sqrt{\alpha N} \leqslant |v| \leqslant \pi\sqrt{\alpha N}\},$$
$$S_2 = \{v : \sqrt[7]{N} \leqslant |v| \leqslant \delta\sqrt{\alpha N}\},$$
$$S_3 = \{v : |v| \leqslant \sqrt[7]{N}\}.$$

In the domain S_1 we have, for $|t| \leqslant \sqrt[8]{N}$,

$$\left|\prod_{k-1}^{N} A_{1k}(v)\right| = \left|\exp\left\{-\alpha N\left(1 - \cos\frac{v}{\sqrt{\alpha N}}\right) + i\alpha N \sin\frac{v}{\sqrt{\alpha N}}\right\}\right| \leqslant$$
$$\leqslant \exp\left\{-2\alpha N \sin^2\frac{\delta}{2}\right\},$$

$$\left|\prod_{k=1}^{N} A_{2k}(v)\right| = \prod_{k=1}^{N}\left(1 + O\left(\frac{\sqrt[8]{N}}{\sqrt{N}}\right)\right) = e^{O(N^{5/8})}.$$

Therefore, for $|t| \leqslant \sqrt[8]{N}$,

$$\int_{S_1}\left|\prod_{k=1}^{N} A_{1k}(v) A_{2k}(v)\right| dv \leqslant$$
$$\leqslant 2\pi\sqrt{\alpha N} \exp\left\{-2\alpha N \sin^2\frac{\delta}{2} + O\left(N^{5/8}\right)\right\} = \bar{o}(e^{-t^2/2}). \quad (7)$$

Let us define

$$\rho = \sqrt{v^2 + t^2}. \quad (8)$$

It is easy to verify that the following relations hold:

$$\ln \prod_{k=1}^{N} A_{1k}(v) = -\frac{v^2}{2} + iv\sqrt{\alpha N} + O\left(\frac{|v|^3}{\sqrt{N}}\right),$$

$$\ln \prod_{k=1}^{N} A_{2k}(v) = \frac{itm_N\sqrt{N}}{\sigma_N} + \frac{\sqrt{\alpha}\,vt}{\sigma_N}\left(\sum_{k=1}^{N} e^{-\alpha N a_k} a_k\right) - \tag{9}$$

$$-\frac{t^2}{2\sigma_N^2}\left[m_N - \left(\frac{1}{N}\sum_{k=1}^{N} e^{-2\alpha N a_k}\right)\right] + O\left(\frac{\rho^3}{\sqrt{N}}\right).$$

These equations are obviously satisfied in the domain $S_2 \cup S_3$ for $|t| < \sqrt[8]{N}$. Using (9) and the inequalities

$$\frac{|v|^3}{\sqrt{N}} = \frac{|v|}{\sqrt{\alpha N}}\sqrt{\alpha}\,v^2 < \delta\sqrt{\alpha}\,v^2, \quad |v| > \sqrt[7]{N}, \quad |t| < \sqrt[8]{N},$$

$$\rho^3 = |v^3|\sqrt{\left(1+\frac{t^2}{v^2}\right)^3} < 3|v|^3,$$

we obtain the estimate

$$\left|\prod_{k=1}^{N} A_{1k}(v)\,A_{2k}(v)\right| < \exp\left\{-\frac{v^2}{2} + K(\delta v^2 + |v|\cdot|t| + t^2)\right\}$$

in the domain S_2. Let us choose δ so that $K\delta < 1/4$. Then noting that $|v| > \sqrt[7]{N}$, $|t| < \sqrt[8]{N}$, we can express the last inequality for sufficiently large N in the form

$$\left|\prod_{k=1}^{N} A_{1k}(v)\,A_{2k}(v)\right| < e^{-c_\delta v^2},$$

where c_δ is a positive constant. From this we see that

$$\int_{S_2}\left|\prod_{k=1}^{N} A_{1k}(v)\,A_{2k}(v)\right| dv = o\,(e^{-t^2/2}) \tag{10}$$

as $n,\ N \to \infty$. Making use of (7) and (10), we obtain from (2)

$$\Psi_N(t) = \frac{1+o(1)}{\sqrt{2\pi}} e^{-\frac{itm_N\sqrt{N}}{\sigma_N}} \times \int_{S_3} e^{-iv\sqrt{\alpha N}} \prod_{k=1}^{N} A_k(v)dv + o(e^{-t^2/2}). \tag{11}$$

Adding the right-hand sides of equations (9), we find

$$\ln \prod_{k=1}^{N} A_k(v) = \frac{itm_N\sqrt{N}}{\sigma_N} + iv\sqrt{\alpha N} - $$
$$- \frac{t^2}{2\sigma_N^2}\left[m_N - \frac{1}{N}\sum_{k=1}^{N} e^{-2\alpha N a_k} - \alpha\left(\sum_{k=1}^{N} a_k e^{-\alpha N a_k}\right)^2\right] -$$
$$- \frac{1}{2}\left[v - \frac{t\sqrt{\alpha}}{\sigma_N}\left(\sum_{k=1}^{N} a_k e^{-\alpha N a_k}\right)\right]^2 + O\left(\frac{\rho^3}{\sqrt{N}}\right). \tag{12}$$

Substituting this expression into (11) and taking into account (1.18), we see that

$$\Psi_N(t) = \frac{1+o(1)}{\sqrt{2\pi}} e^{-t^2/2} \int_{-\sqrt[7]{N}}^{\sqrt[7]{N}} \left[1 + O\left(\frac{\rho^3}{\sqrt{N}}\right)\right] \times$$
$$\times \exp\left\{-\frac{1}{2}\left[v - \frac{t\sqrt{\alpha}}{\sigma_N}\left(\sum_{k=1}^{N} a_k e^{-\alpha N a_k}\right)\right]^2\right\} dv \to e^{-t^2/2}$$

uniformly with respect to $|t| < \sqrt[8]{N}$ as $n, N \to \infty$. The theorem is proved.

From the above theorem we get immediately the following integral limit theorem:

Theorem 2. *In the central domain given by* (1.12),

$$\mathbf{P}\left\{\frac{\mu_0 - Nm_N}{\sigma_N\sqrt{N}} < x\right\} \to \frac{1}{\sqrt{2\pi}} \int_{-\infty}^{x} e^{-u^2/2} du$$

as $n, N \to \infty$, *where* m_N, σ_N *are defined by* (1.17) *and* (1.18).

To prove the local theorem, we shall need two lemmas. It follows from (12) that

$$\left|\prod_{k=1}^{N} A_k(v)\right| \leqslant \exp\left(-\frac{1}{2} Q(t, v) + K \frac{\rho^3}{\sqrt{N}}\right), \tag{13}$$

where $K>0$ is some constant,

$$\begin{gathered} Q(u_1, u_2) = \sum_{i,j=1}^{2} q_{ij} u_i u_j, \\ q_{11} = \sigma_N^2 + \alpha\left(\sum_{k=1}^{N} a_k e^{-\alpha N a_k}\right)^2, \quad q_{22} = 1, \\ q_{12} = q_{21} = -\sqrt{\alpha} \sum_{k=1}^{N} a_k e^{-\alpha N a_k}. \end{gathered} \tag{14}$$

Lemma 1. *There exists a constant $\delta>0$ such that in the domain $\rho=\sqrt{v^2+t^2}<\delta\sqrt{N}$ we have the inequality*

$$\left|\prod_{k=1}^{N} A_k(v)\right| < e^{-c_\delta \rho^2},$$

where c_δ is a positive constant depending on δ.

PROOF. According to Theorem 1.3, we have $q_{11} \geqslant \sigma_N^2 \geqslant \varepsilon > 0$ and

$$\begin{vmatrix} q_{11} & q_{12} \\ q_{21} & q_{22} \end{vmatrix} = \begin{vmatrix} \sigma_N^2 + q_{12}^2 & q_{12} \\ q_{12} & 1 \end{vmatrix} = \sigma_N^2 \geqslant \varepsilon > 0.$$

Thus, the quadratic form $Q(u_1, u_2)$ is positive-definite. Therefore,

$$\min_{\rho=1} Q(t, v) = c(\alpha, a_1, \ldots, a_N) > 0$$

for any $\alpha, a_1, \ldots, a_N$, varying in the central domain (1.12). Since $c(\alpha, a_1, \ldots, a_N)$ is continuous with respect to

$\alpha, a_1, \ldots, a_N$, there exists under conditions (1.12) a positive constant $\tilde{c}$ such that we have $c(\alpha, a_1, \ldots, a_N) \geqslant \tilde{c}$. From this we get $Q(t, v) \geqslant \tilde{c}\rho^2$. Since $\rho^3/\sqrt{N} \leqslant \delta\rho^2$ (under the conditions of the lemma), the assertion of the lemma with $c_\delta = \tilde{c} - K\delta$ follows from (13) if $K\delta < \tilde{c}$.

Let us estimate now the product in (2) for all admissible values t and v (i.e., $|t| \leqslant \pi\sigma_N\sqrt{N}$, $|v| \leqslant \pi\sqrt{\alpha N}$). Using (3), one can verify that

$$\prod_{k=1}^{N} A_k(v) = \prod_{k=1}^{N} I\left(\frac{t}{\sigma_N \sqrt{N}}, \frac{v}{\sqrt{\alpha N}}, \alpha N a_k\right), \tag{15}$$

where

$$I(x, y, \beta) = e^{\beta(e^{iy}-1)} + (e^{ix} - 1)\, e^{-\beta}. \tag{16}$$

Lemma 2. *For any* $\delta > 0$, *there exists a constant* q *in* $(0, 1)$ *such that*

$$\left|\prod_{k=1}^{N} A_k(v)\right| \leqslant q^N,$$

if either $|v| \geqslant \delta\sqrt{\alpha N}$ *or* $|t| \geqslant \delta\sigma_N\sqrt{N}$.

PROOF. Let us define

$$x = \frac{t}{\sigma_N \sqrt{N}}, \quad y = \frac{v}{\sqrt{\alpha N}} \quad (|x| \leqslant \pi, |y| \leqslant \pi).$$

We show first that

$$|I(x, y, \beta)| \leqslant 1 \tag{17}$$

or

$$e^{\beta} - \left|e^{\beta e^{iy}} + e^{ix} - 1\right| \geqslant 0.$$

For the left-hand side of this inequality, we have

$$e^{\beta} - \left|e^{\beta e^{iy}} + e^{ix} - 1\right| \geqslant e^{\beta} - \left|e^{\beta e^{iy}} - 1\right| - \left|e^{ix}\right| =$$
$$= e^{\beta} - 1 - \left|e^{\beta e^{iy}} - 1\right| \geqslant 0.$$

The first inequality in this chain becomes equality for $\beta > 0$ if

$$x = \arg\left[e^{\beta e^{iy}} - 1\right],$$

but the second inequality becomes equality only if $y = 0$. Therefore, (17) becomes equality for $\beta > 0$ only if $x = y = 0$. From this fact and continuity of $I(x, y, \beta)$ it follows that for any $\delta > 0$ there exists an $\varepsilon > 0$ such that

$$|I(x, y, \beta)| < 1 - \varepsilon \tag{18}$$

for any $\beta > \delta$ if either $|x| > \delta$ or $|y| > \delta$. The estimate (1.21) holds for the number of probabilities a_k satisfying (1.20). Therefore, the number N_δ of values of k for which the inequality $\alpha N a_k \geqslant \delta > 0$ is satisfied is not less than $c_0 N$, where c_0 is a positive constant. Using (15), (17), and (18) we arrive at

$$\left|\prod_{k=1}^{N} A_k(v)\right| \leqslant (1 - \varepsilon)^{c_0 N},$$

thus proving the lemma.

Theorem 3. *In the central domain given by* (1.12), *as* $n, N \to \infty$,

$$\mathbf{P}\{\mu_0 = k\} = \frac{1}{\sqrt{2\pi N}\sigma_N} e^{-u^2/2}(1 + o(1))$$

uniformly with respect to $\alpha, a_1, \ldots, a_N$ *and uniformly with respect to* $k = Nm_N + u\sigma_N\sqrt{N}$, *where* $|u| \leqslant C < \infty$. *The variables* m_N, σ_N *are defined by* (1.17) *and* (1.18).

Proof. The probability distribution $\mathbf{P}\{\mu_0 = k\}$ is connected with the characteristic function $\Psi_N(t)$ defined by (1) as follows:

$$\mathbf{P}\{\mu_0 = k\} = \frac{1}{2\pi}\int_{-\pi}^{\pi} e^{-it(k - Nm_N)}\, \Psi_N\left(t\sigma_N\sqrt{N}\right) dt. \tag{19}$$

Let us set $\tau = t\sigma_N\sqrt{N}$ and express (19) as $\mathbf{P}\{\mu_0 = k\} = I_1 + I_2$, where

$$I_1 = \frac{1}{2\pi\sigma_N\sqrt{N}} \int\limits_{-\sqrt[8]{N}}^{\sqrt[8]{N}} e^{-iu\tau}\Psi_N(\tau)\, d\tau$$

$$I_2 = \frac{1}{2\pi\sigma_N\sqrt{N}} \int\limits_{\sqrt[8]{N} \leqslant |\tau| \leqslant \pi\sigma_N\sqrt{N}} e^{-iu\tau}\Psi_N(\tau)\, d\tau. \qquad (20)$$

Making use of Theorem 1, we see that

$$I_1 = \frac{1+o(1)}{2\pi\sqrt{N}\sigma_N} \int\limits_{-\sqrt[8]{N}}^{\sqrt[8]{N}} e^{-iu\tau - \frac{\tau^2}{2}}\, d\tau =$$

$$= \frac{1}{\sqrt{2\pi N}\sigma_N} e^{-u^2/2}(1+o(1)) \qquad (21)$$

as $n,\ N \to \infty$.

In the second integral in (20), we substitute for $\Psi_N(\tau)$ its value given by (2):

$$I_2 = \frac{1+o(1)}{(2\pi)^{3/2}\sigma_N\sqrt{N}} \times$$

$$\times \iint\limits_S \exp\left(-iut - it\frac{m_N\sqrt{N}}{\sigma_N} - iv\sqrt{\alpha N}\right) \prod_{k=1}^{N} A_k(v)\, dv\, dt, \qquad (22)$$

where the $A_k(v)$ are defined by (3) and

$$S = \{(t, v):\ \sqrt[8]{N} \leqslant |t| \leqslant \pi\sigma_N\sqrt{N},\ |v| \leqslant \pi\sqrt{\alpha N}\}.$$

Let us define

$$S_1 = \{(t, v): \rho = \sqrt{t^2 + v^2} \leqslant \delta\sqrt{N}\} \cap S,$$

$$S_2 = S \setminus S_1.$$

We partition the integration domain in (22) into the two subdomains S_1 and S_2. Then $I_2 = I_{S_1} + I_{S_2}$, where I_{S_i} is given by (22) with S replaced by S_i. To estimate I_{S_1} we make use of Lemma 1:

$$|I_{S_1}| \leqslant \frac{K}{\sqrt{N}} \iint\limits_{S_1} e^{-c_\delta \rho^2} dt\, dv \leqslant \frac{K}{\sqrt{N}} \iint\limits_{\rho \geqslant \sqrt[8]{N}} e^{-c_\delta \rho^2} dt\, dv,$$

where K is a positive constant. The right-hand side of the last inequality approaches zero as $N \to \infty$. Therefore, as $n, N \to \infty$.

$$I_{S_1} = o\left(\frac{1}{\sqrt{N}}\right). \tag{23}$$

To estimate I_{S_2}, we use Lemma 2 and formula (22):

$$|I_{S_2}| \leqslant \frac{K}{\sqrt{N}} \iint\limits_{S_2} q^N dv\, dt = O\left(N^{3/2} q^N\right). \tag{24}$$

The assertion of the theorem follows from (19)-(24).

§5. Multivariate Normal Limit Theorems

Let us consider a random vector $\mu_{r_1}, \mu_{r_2}, \ldots, \mu_{r_s}$. Denote by (x, y) the scalar product of two vectors $(x_1, \ldots, x_s)$ and $(y_1, \ldots, y_s)$. Let

$$\Psi_N(t) = \mathbf{M} e^{i(t, \xi)}, \tag{1}$$

where the vectors $t = (t_1, t_2, \ldots, t_s)$, $\xi = (\xi_1, \ldots, \xi_s)$, $\xi_k = \left(\mu_{r_k} - N\widetilde{m}_{r_k}\right) N^{-1/2}$, and $\widetilde{m}_{r_k}$ are as defined in Theorem 1.5.

Theorem 1. *In the central domain defined by* (1.12), *as* $n, N \to \infty$,

$$\Psi_N(t) = \exp\{-(\|\tilde{\sigma}_{r_i r_j}\| t, t)\}(1 + o(1)),$$

(where $\Psi_N(t)$ *is the characteristic function defined by* (1) *and the* $\tilde{\sigma}_{r_i r_j}$ *are defined in Theorem* 1.5) *uniformly with respect to* $|t_k| \leqslant \sqrt[8]{N}$, $k = 1, 2, \ldots, s$.

PROOF. Using (1.6) as in the proof of Theorem 4.1, we get

$$\Psi_N(t) = \frac{1 + o(1)}{\sqrt{2\pi}} e^{-i(t,\tilde{m})\sqrt{N}} \int_{-\pi\sqrt{\alpha N}}^{\pi\sqrt{\alpha N}} e^{-iv\sqrt{\alpha N}} \prod_{k=1}^{N} A_{1k}(v) A_{2k}(v)\, dv, \tag{2}$$

where $\tilde{m} = (m_{r_1}, \ldots, m_{r_s})$,

$$A_{1k}(v) = \exp\left\{\alpha N a_k \left(e^{\frac{iv}{\sqrt{\alpha N}}} - 1\right)\right\}, \tag{3}$$

$$A_{2k}(v) = 1 + \sum_{l=1}^{s} p_{r_l,k} e^{\frac{i r_l v}{\sqrt{\alpha N}}} \left(e^{\frac{i t_l}{\sqrt{N}}} - 1\right) \exp\left\{-\alpha N a_k \left(e^{\frac{iv}{\sqrt{\alpha N}}} - 1\right)\right\}, \tag{4}$$

where $p_{r,k}$ are defined by (1.22). The remainder of the proof of the theorem follows the proof of Theorem 1 if in place of (4.9) and (4.12) we use the formulas

$$\ln \prod_{k=1}^{N} A_{1k}(v) = -\frac{v^2}{2} + iv\sqrt{\alpha N} + O\left(\frac{|v|^3}{\sqrt{N}}\right),$$

$$\ln \prod_{k=1}^{N} A_{2k}(v) = i\sqrt{N}(t, \tilde{m}) - \frac{1}{2} \sum_{l=1}^{s} m_{r_l} t_l^2 +$$
$$+ \frac{v}{\sqrt{\alpha}} \sum_{l=1}^{s} t_l \left(\frac{1}{N} \sum_{k=1}^{N} p_{r_l,k} (\alpha N a_k - r_l)\right) +$$
$$+ \frac{1}{2} \sum_{l_1, l_2 = 1}^{s} t_{l_1} t_{l_2} \left(\frac{1}{N} \sum_{k=1}^{N} p_{r_{l_1},k} p_{r_{l_2},k}\right) + O\left(\frac{\rho^3}{\sqrt{N}}\right),$$

$$\ln \prod_{k=1}^{N} A_{1k}(v)A_{2k}(v) = iv\sqrt{\alpha N} + i\sqrt{N}\,(t, \tilde{m}) - \frac{1}{2}\left(\|\tilde{\sigma}_{r_i r_j}\|t, t\right) -$$

$$-\frac{1}{2}\left[v - \frac{1}{\sqrt{\alpha}}\sum_{l=1}^{s}\frac{t_l}{N}\sum_{k=1}^{N} p_{r_l,k}(\alpha N a_k - r_l)\right]^2 + O\left(\frac{\rho^3}{\sqrt{N}}\right), \quad (5)$$

where $\rho = \sqrt{t_1^2 + t_2^2 + \ldots + t_s^2 + v^2}$.

A corollary of Theorem 1 is

Theorem 2., *In the central domain defined by* (1.12), *we have, uniformly with respect to* $x_1, x_2, \ldots, x_s$ *as* $n, N \to \infty$,

$$\left|\mathbf{P}\left\{\frac{\mu_{r_l} - N\tilde{m}_{r_l}}{\sqrt{N}} < x_l, l = 1, \ldots, s\right\} - \Phi(x_1, \ldots, x_s; \|\tilde{\sigma}_{r_i r_j}\|)\right| \to 0,$$

where $\Phi(x_1, \ldots, x_s; \|\tilde{\sigma}_{r_i r_j}\|)$ *is the s-dimensional normal law with mean zero and covariance matrix* $\|\tilde{\sigma}_{r_i r_j}\|$, *the variables* $\tilde{m}_r$, $\tilde{\sigma}_{r_i r_j}$ *being defined by the formulas in Theorem 1.5.*

The local normal theorem will be used in Chapter 5 in developing optimality criteria. It is more convenient to state and prove such a theorem in the case where the assumptions (1.23) are satisfied. Let us prove first the generalization of Lemmas 4.1 and 4.2.

Lemma 1. *If the conditions given in* (1.23) *are satisfied, there exists a constant* $\delta > 0$ *such that in the domain* $\rho = \sqrt{t_1^2 + \ldots + t_s^2 + v^2} < \delta\sqrt{N}$ *we have the inequality*

$$\left|\prod_{k=1}^{n} A_{1k}(v)\,A_{2k}(v)\right| < e^{-c_\delta \rho^2},$$

where c_δ. *is some positive constant depending on* δ *and where* $A_{1k}(v)$ *and* $A_{2k}(v)$ *are defined by* (3) *and* (4).

PROOF. It follows from (5) that

$$\left|\prod_{k=1}^{N} A_{1k}(v)\,A_{2k}(v)\right| \leqslant \exp\left(-\frac{1}{2}Q(t, v) + K\frac{\rho^3}{N}\right),$$

where K is some constant

$$Q(t, v) = \left(\|q_{ij}\| t, t\right) + 2\sum_{i=1}^{s} t_i v q_i + v^2, \tag{6}$$

$\|q_{ij}\|$ is the $s \times s$ matrix defined by $q_{ij} = \tilde{\sigma}_{r_i r_j} + q_i q_j$, where

$$q_i = -\frac{1}{N\sqrt{\alpha}} \sum_{k=1}^{N} p_{r_i,k} (\alpha N \bar{a}_k - r_i), \quad i = 1, 2, \ldots, s.$$

The quadratic form given by (6) is positive-definite if its principal minors are positive. All these minors are of the same form as the determinant of the matrix itself.

To prove that

$$\begin{vmatrix} 1 & q_1 & q_2 & \ldots & q_s \\ q_1 & \tilde{\sigma}_{r_1 r_1} + q_1^2 & \tilde{\sigma}_{r_1 r_2} + q_1 q_2 & \ldots & \tilde{\sigma}_{r_1 r_s} + q_1 q_s \\ \cdot & \cdot & \cdot & \cdot & \cdot \\ q_s & \tilde{\sigma}_{r_s r_1} + q_s q_1 & \tilde{\sigma}_{r_s r_2} + q_s q_2 & \ldots & \tilde{\sigma}_{r_s r_s} + q_s^2 \end{vmatrix} = \left|\|\tilde{\sigma}_{r_i r_j}\|\right|, \tag{7}$$

we multiply the first row by q_i (for ($i = 1, 2, \ldots, s$) and subtract it from the $(i+1)$st row. By Theorem 1.6, we have $\tilde{\sigma}_{r_i r_j} \to \sigma_{r_i r_j}$ as $N \to \infty$. Therefore, the right-hand side of (7) tends to the determinant B^2, which can be found by formula (2.2.5). By the corollary of Lemma 2.2.1,

$$B^2 \geqslant \varepsilon > 0 \quad \text{for } \alpha \in [\alpha_0, \alpha_1].$$

From this, as in the proof of Lemma 4.1, the assertion of the lemma follows.

Lemma 2. *For any* $\delta > 0$, *there exists a constant* q *in* (0, 1) *such that*

$$\left|\prod_{k=1}^{N} A_{1k}(v) A_{2k}(v)\right| \leqslant q^N, \tag{8}$$

if any of the inequaltities $|v|>\delta\sqrt{\alpha N}$, $|t_1|>\delta\sqrt{N}$, $|t_s|>\delta\sqrt{N}$ *hold.*

PROOF. It is easy to verify that the product between the absolute-value bars in the left-hand side of (8) is equal to

$$\prod_{k=1}^{N} I\left(\frac{t}{\sqrt{N}}, \frac{v}{\sqrt{\alpha N}}, \alpha N a_k\right),$$

where

$$I(x, y, \beta) = e^{\beta(e^{iy}-1)} + \sum_{l=1}^{s} \frac{\beta^{r_l}}{r_l!} e^{-\beta} (e^{ix_l} - 1) e^{iyr_l}.$$

As in the proof of Lemma 4.2, it suffices to show here that we have the conditional inequality

$$|I(x, y, \beta)| \leqslant 1 \tag{9}$$

with equality holding in the domain

$$\{\beta>0, \ |y|\leqslant\pi, \ |x_l|\leqslant\pi, \ l=1, \ldots, s\}$$

only at $x_1=\ldots=x_s=y=0$.

To prove (9), we note that

$$e^{\beta} - |e^{\beta} I(x, y, \beta)| =$$

$$= e^{\beta} - \left| e^{\beta e^{iy}} - \sum_{l=1}^{s} \frac{\beta^{r_l}}{r_l!} e^{iyr_l} + \sum_{l=1}^{s} \frac{\beta^{r_l}}{r_l!} e^{i(yr_l+x_l)} \right| \geqslant$$

$$\geqslant e^{\beta} - \left| e^{\beta e^{iy}} - \sum_{l=1}^{s} \frac{\beta^{r_l}}{r_l!} e^{iyr_l} \right| - \sum_{l=1}^{s} \left| \frac{\beta^{r_l}}{r_l!} e^{i(yr_l+x_l)} \right| =$$

$$= \sum_{k=0}^{\infty} \frac{\beta^k}{k!} - \left| \sum_{k=0}^{\infty} \frac{\beta^k}{k!} e^{iyk} - \sum_{l=1}^{s} \frac{\beta^{r_l}}{r_l!} e^{iyr_l} \right| - \sum_{l=1}^{\infty} \frac{\beta^{r_l}}{r_l!} =$$

$$= \sum_{\substack{k\neq r_l \\ l=1,\ldots,s}} \frac{\beta^k}{k!} - \left| \sum_{\substack{k\neq r_l \\ l=1,\ldots,s}} \frac{\beta^k}{k!} e^{iyk} \right| \geqslant 0. \tag{10}$$

Since equality holds in the conditional inequality $\left|\sum_{l=1}^{s} z_l\right| \leqslant \sum_{l=1}^{s} |z_l|$ for $\arg z_1 = \ldots = \arg z_s$ only if $\beta > 0$, the first inequality in (10) becomes equality only for one set of $x_1, x_2, \ldots, x_s$ depending on y. The second inequality in (10) becomes equality only if $y=0$. The proof of the lemma is readily completed by arguing as in Lemma 4.2.

Theorem 3. *If conditions* (1.23) *are satisfied, then, as* n, $N \to \infty$, *we have uniformly with respect to* $|u_i| \leqslant C < \infty$, *where* $u_i = (k_i - N\tilde{m}_{r_i})/\sqrt{N}$, *and* $\alpha = n/N \in [\alpha_0, \alpha_1]$, *where* $0 < \alpha_0 \leqslant \alpha_1 < \infty$,

$$\mathbf{P}\{\mu_{r_1} = k_1, \ldots, \mu_{r_s} = k_s\} = \frac{1+o(1)}{(2\pi N)^{s/2}B} \exp\left(-\frac{1}{2B^2}\sum_{i,j=1}^{s} B_{ij}u_iu_j\right),$$

where $p_{r_i} = \frac{\alpha^{r_i}}{r_i!}e^{-\alpha}$, B^2, B_{ij} *are defined by* (2.2.5), (2.2.8), *and* (2.2.9) *and* $\tilde{m}_r$ *is defined by* (1.24).

This theorem is proved with the aid of Lemmas 1 and 2 and Theorem 1 in a manner similar to the proof of Theorem 4.3.

We need to note only that, in decomposing the s-dimensional integral that takes the place of (4.19) into the sum I_1+I_2, we must integrate in the case of I_1 over the domain

$$T = \{t: |t_1| < \sqrt[8]{N}, \ldots, |t_s| < \sqrt[8]{N}\}.$$

and in the case of I_2 over the domain $\{t: \max|t_i| < \pi\sqrt{N}\} \setminus T$. In determining the domains S, S_1, and S_2 we must take

$$\rho = \sqrt{t_1^2 + \ldots + t_s^2 + v^2}.$$

§6. Further Results. References

The generating functions (1.1) and (1.2) were obtained by Sevast'yanov and Chistyakov [47]. The Poisson limit theorem for the sum of indicators was proved by Sevast'yanov in [46]; the treatment of this problem given in Section 2 follows that of [46]. The Poisson limit theorem is used by Sevast'yanov for proving the theorems in Section 3.

Some limit distributions for μ_0 different from Poisson distributions were obtained by Chistyakov. In [56] he proved the following statement:

If $\frac{n}{N^2} \to 0$ as $n, N \to \infty$, if $a_1 \leqslant a_2 \leqslant \ldots \leqslant a_N$, and if

$$\mathbf{M}\mu_0 = \sum_{k=1}^{N} (1 - a_k)^n \to m, \qquad (1 - a_k)^n \to \gamma_k, \qquad 0 \leqslant \gamma_k \leqslant 1,$$

where m and γ_k are constants, then

$$\mathbf{M}x^{\mu_0} \to e^{\lambda(x-1)} \prod_{k=1}^{\infty} (1 - \gamma_k + \gamma_k x), \ \lambda = m - \sum_{k=1}^{\infty} \gamma_k \geqslant 0. \quad (1)$$

By using the method of moments, it is possible to prove the above statement without assuming that $\frac{n}{N^2} \to 0$. The proof is similar to that of Theorems 2.1 and 3.3. We need to show that the asymptotic behavior of the factorial moments μ_0 is determined only by those a_k satisfying the inequality $a_k \leqslant n^{-3/4}$. The generalization of (1) to the case in which μ_0 is the number of nonappearing s-chains in a polynomial scheme, is due to Kolchin and Chistyakov [35].

Theorem 4.2 on asymptotic normality of μ_0 was proved in Chistyakov [55]. A similar theorem is formulated by Kitabatake [84] under stronger restrictions on a_k than those given by (1.12). Holst [82] extended Theorem 4.2 to the left-hand intermediate domain; he showed that the normality of μ_0 remains if in (1.12) the condition $0 < \alpha_0 \leqslant \frac{n}{N} \leqslant \alpha_1 < \infty$ is replaced with

$$\frac{n}{N} \to 0, \qquad \frac{n^2}{N} \to \infty.$$

The multivariate normal integral theorem was proved and the corresponding local theorem is formulated in Viktorova and Chistyakov [12]. The complete proof of the local theorem for $\mu_{r_1}, \mu_{r_2}, \ldots, \mu_{r_s}$ was first given by Viktorova [10].

T. Yu. Popova [40] studied the scheme of independent allocation of n_1 particles of the first type and n_2 particles of the second type. A particle of the kth type gets into the ith cell with probability $a_i(k)$, where $k=1, 2$, and $i=1, 2, \ldots, N$. For such a scheme, a Poisson limit theorem and a normal multivariate limit theorem are proved for the number of cells containing no particles or containing no particles of a certain type.

Holst in [81] examined the following problem: Suppose that the cell numbered k is assigned the number c_k, where $k=1, 2, \ldots, N$, in a scheme with N cells and n particles. Denote by η_n the sum of those c_k, corresponding to cells with no more than r particles. We have $\eta_n=\mu_0$ for $c_1=\ldots=c_N=1$, $r=0$. Holst proved the asymptotic normality of the vector

$$\eta_{n_1}, \eta_{n_1+n_2}, \ldots, \eta_{n_1+\ldots+n_d},$$

as $N, n_1, \ldots, n_d \to \infty$. Rosen [94] considered this problem for the special case, $r=0$, using results on the sums of dependent variables obtained by him in [93]. The results related to the "sequential occupancy" in problems involving the sums η_n are given in Rosen [95].

A lattice-filling scheme with regard to the arrangement of particles was investigated by Hafner [76], [77], [78]. The scheme involving an infinite number of cells is a special case. Hafner extends some well-known results to more general cases. Schemes involving an infinite number of cells are also discussed in Darling [69] and Karlin [83].

Chapter IV

CONVERGENCE TO RANDOM PROCESSES

§1. The Statement of the Problem

In this chapter, we consider the number of empty boxes $\mu_0(n, N)$ as a function of the number n of shots thrown, i.e., as a process depending on "time" n. Let us assume that the shots are thrown independently and successively into N boxes. The probability that any one shot will get into any one box is equal to $1/N$. Therefore, the random variables $\mu_0(1, N)$, $\mu_0(2, N), \dots, \mu_0(n, N), \dots$ are all defined on the same probability space. Thus we may consider the joint distribution of these random variables.

In the previous chapters, we proved limit theorems for $\mu_0(n, N)$ as $n, N \to \infty$. We obtained a Poisson distribution for small and large values of the ratio n/N in the limit and a normal distribution for intermediate values of n/N. Thus we arrive at three basic types of convergence of the process $\mu_0(n, N)$ to a limiting process.

If $n, N \to \infty$ in the left-hand 0-domain, i.e., if the values of

$\theta = \frac{n^2}{2N}$ are bounded, then $\mu_0(n, N) - (N - n)$ converges in the limit to a Poisson process $\eta(\theta)$ with $\mathbf{M}\eta(\theta) = \theta$ (see Section 3).

If $n, N \to \infty$ in the right-hand 0-domain, i.e., if $\theta = Ne^{-n/N}$ is bounded, then $\mu_0(n, N)$ converges to a Poisson process $\eta(\theta)$ with $\mathbf{M}\{\eta(\theta_2) - \eta(\theta_1)\} = \theta_2 - \theta_1$, where $\theta_2 > \theta_1$ (see Section 5).

Finally, if $n, N \to \infty$ in the central domain, i.e., if $\theta = \frac{n}{N}$ is bounded, $\xi(\theta) = \frac{\mu_0(n, N) - \mathbf{M}\mu_0(n, N)}{\sqrt{N}}$ converges to a Gaussian process $\eta(\theta)$ with $e^{-\theta'}(1 - (1 + \theta)e^{-\theta})$, where $\theta' \geqslant \theta$ (see Section 4).

Thus a particular time θ is introduced in each domain of variation of n and N. In the left-hand domain $\theta = \frac{n^2}{2N}$; in the central domain $\theta = n/N$; in the right-hand domain $\theta = Ne^{-n/N}$.

The limit behavior of the process $\mu_0(n, N)$ is of a more complicated nature in intermediate domains. In each intermediate domain, we need to introduce "local times," associated with some "zero values" of $\alpha_0 = \frac{n_0}{N} \to 0$ or $\alpha_0 \to \infty$ (see Section 6).

Similar results are also obtained for $\mu_r(n, N)$ in an equiprobable polynomial scheme (see Section 7).

§2. Generating Functions of Multidimensional Distributions

Let us consider an equiprobable scheme. We denote the number of empty boxes $\mu_0(n, N)$ by $\mu_0(n)$, emphasizing the fact that this number depends on the number of shots n. In this chapter, our main analytical tool is the multivariate generating functions of $\mu_0(n_1)$, $\mu_0(n_1+n_2), \ldots, \mu_0(n_1+n_2+\ldots+n_t)$, where the n_i are nonnegative integers.

Let

$$F_{n_1 n_2 \ldots n_t}(x_1, \ldots, x_t) = \mathbf{M} x_1^{\mu_0(n_1)} x_2^{\mu_0(n_1+n_2)} \ldots x_t^{\mu_0(n_1+\ldots+n_t)} \quad (1)$$

be the generating function of the joint distribution of the random variables $\mu_0(n_1)$, $\mu_0(n_1+n_2)$,..., $\mu_0(n_1+...+n_t)$. We introduce the generating function

$$\Phi_N(z, x) = \Phi_N(z_1,..., z_t; x_1,..., x_t) =$$
$$= \sum_{n_1,...,n_t}^{\infty} F_{n_1...n_t}(x_1,..., x_t) \prod_{l=1}^{t} \frac{(Nz_l)}{n_l!}. \quad (2)$$

Theorem 1. *The generating function* $\Phi_N(z; x)$ *is of the following form:*

$$\Phi_N(z; x) = [(e^{z_1}-1)\, e^{z_2+...+z_t} + (e^{z_2}-1)\, e^{z_3+...+z_t} x_1 + ...$$
$$... + (e^{z_{t-1}}-1)\, e^{z_t} x_1 x_2 ... x_{t-2} + (e^{z_t}-1)\, x_1 x_2 ... x_{t-1} +$$
$$+ x_1 x_2 ... x_t]^N. \quad (3)$$

PROOF. Let us divide the N boxes into two groups. Let there be N_1 boxes in the first group and N_2 boxes in the second group, where $N_1+N_2=N$. Suppose that, out of n_l shots, n_{1l} shots get into the first group and n_{2l} into the second group, where $n_l=n_{1l}+n_{2l}$. We denote by $\mu_{10}(n_{11}+n_{12}+...+n_{1l})$ and $\mu_{20}(n_{21}+n_{22}+...+n_{2l})$ the numbers of boxes in the first and the second group respectively after $n_1+n_2+...+n_l$ shots are thrown into the boxes. By the law of composition of probabilities, we have

$$\mathbf{P}\{\mu_0(n_1+...+n_l) = k_l,\ l = 1,..., t\} =$$
$$= \sum_{D_{n_1...n_t}} \sum_{E_{k_1...k_t}} \prod_{l=1}^{t} \frac{n_l!}{n_{1l}!\, n_{2l}!} \cdot \frac{N_1^{n_{1l}} N_2^{n_{2l}}}{N^{n_l}} \times$$
$$\times \mathbf{P}\{\mu_{10}(n_{11}+...+n_{1l}) = k_{1l},\ l = 1, ..., t\} \times$$
$$\times \mathbf{P}\{\mu_{20}(n_{21}+...+n_{2l}) = k_{2l},\ l = 1,..., t\}, \quad (4)$$

where the summation domains are defined as follows:

$$D_{n_1...n_t} = \{n_{1l}, n_{2l},\ l = 1,..., t:\ n_{1l}+n_{2l} = n_l,\ l = 1,..., t\},$$
$$E_{k_1...k_t} = \{k_{1l}, k_{2l},\ l = 1,..., t:\ k_{1l}+k_{2l} = k_l,\ l = 1,..., t\}.$$

In deriving formula (4), we assume first that, when n_l shots are thrown, they distribute themselves in groups of n_{1l} and n_{2l} shots to each group (for each l these numbers are distributed according to the binomial law

$$C_{n_l}^{n_{1l}}\left(\frac{N_1}{N}\right)^{n_{1l}}\left(1-\frac{N_1}{N}\right)^{n_{2l}},$$

and, for different l, the random variables n_{il} are independent). Next we use the fact that the conditional joint distribution $\mu_0(n_1+n_2+\ldots+n_l)$ for fixed n_{il} is defined as the convolution of the distributions $\mu_{10}(n_{11}+\ldots+n_{1l})$ and $\mu_{20}(n_{21}+\ldots+n_{2l})$ since μ_{10} and μ_{20} are independent and $\mu_0=\mu_{10}+\mu_{20}$. We multiply both sides of (4) by $\prod_{l=1}^{t}\left(\frac{z_l^{n_l}x_l^{k_l}}{n_l!}\right)$ and sum them with respect to all n_l, k_l. We obtain

$$\Phi_N(z;\ x)=\Phi_{N_1}(z;\ x)\,\Phi_{N_2}(z;\ x),$$

from which we get

$$\Phi_N(z;\ x)=[\Phi(z;\ x)]^N,$$

where $\Phi(z;\ x)=\Phi_1(z;\ x)$. Let us find $\Phi(z;\ x)$. For $N=1$, it is easy to find the generating function $F_{n_1\ldots n_t}(x_1,\ \ldots,\ x_t)$ defined by (1). We group all vectors $\mathbf{n}=(n_1,\ n_2,\ \ldots,\ n_t)$, where $n_i\geqslant 0$, into sets K_0, K_1, $\ldots$, K_t. If $1\leqslant r<t$, we assign $\mathbf{n}$ to K_r if $n_1=\ldots=n_r=0$ and $n_{r+1}>0$. We assign $\mathbf{n}$ to K_0 if $n_1>0$, We assign $\mathbf{n}$ to K_t if $n_1=n_2=\ldots=n_t=0$. If $\mathbf{n}\in K_0$, then

$$\mathbf{P}\{\mu_0(n_1)=\ldots=\mu_0(n_1+\ldots+n_t)=0\}=1;$$

if $\mathbf{n}\in K_r$, where $1\leqslant r<t$, then

$$\mathbf{P}\{\mu_0(n_1)=\ldots=\mu_0(n_1+\ldots+n_r)=1,$$
$$\mu_0(n_1+\ldots+n_{r+1})=\ldots=\mu_0(n_1+\ldots+n_t)=0\}=1;$$

if $\mathbf{n} \in K_t$, then

$$\mathbf{P}\{\mu_0(n_1) = \ldots = \mu_0(n_1 + \ldots + n_t) = 1\} = 1.$$

From this we have

$$F_{\mathbf{n}}(x) = 1, \quad \mathbf{n} \in K_0,$$
$$F_{\mathbf{n}}(x) = x_1 \ldots x_r, \ \mathbf{n} \in K_r, \ 1 \leqslant r \leqslant t. \tag{5}$$

Substituting (5) into (2), we get

$$\Phi(z; x) = \sum_{\mathbf{n} \in K_0} \frac{z_1^{n_1}}{n_1!} \cdot \frac{z_2^{n_2}}{n_2!} \cdots \frac{z_t^{n_t}}{n_t!} + \sum_{r=1}^{t-1} x_1 \ldots x_r \sum_{\mathbf{n} \in K_r} \frac{z_1^{n_1}}{n_1!} \cdots \frac{z_t^{n_t}}{n_t!} +$$

$$+ x_1 \ldots x_t \sum_{\mathbf{n} \in K_t} \frac{z_1^{n_1}}{n_1!} \cdots \frac{z_t^{n_t}}{n_t!} = (e^{z_1} - 1) e^{z_2 + \ldots + z_t} +$$

$$+ \sum_{k=1}^{t-2} \left(e^{z_{k+1}} - 1\right) e^{z_{k+2} + \ldots + z_t} x_1 x_2 \ldots x_k +$$

$$+ \left(e^{z_t} - 1\right) x_1 \ldots x_{t-1} + x_1 x_2 \ldots x_t,$$

thus completing the proof.

§3. Convergence to a Poisson Process in the Left-Hand Domain

In this section, we assume that $n, N \to \infty$ in such a way that $\frac{n^2}{2N} = \theta$ is bounded. We take the variable θ as a new time parameter. Let us prove that under these conditions the process $\mu_0(n) - (N - n)$ converges to a Poisson process $\eta(\theta)$ with $\mathbf{M}\eta(\theta) = \theta$.

Theorem 1. *Suppose that* $n_l, N \longrightarrow \infty$ *in such a way that* $(n_1 + \ldots + n_l)^2 / 2N \to \theta_l$, *where* $0 = \theta_0 < \theta_1 < \ldots < \theta_t < \infty$.

Then for any $t, k_1, \ldots, k_t,$

$$\mathbf{P}\{\mu_0(n_1) - N + n_1 = k_1, \quad \mu_0(n_1 + \ldots + n_l) - \\ - \mu_0(n_1 + \ldots + n_{l-1}) + n_l = k_l, \; l = 2, \ldots, t\} \to \\ \to \prod_{l=1}^{t} \frac{(\theta_l - \theta_{l-1})^{k_l}}{k_l!} e^{-(\theta_l - \theta_{l-1})}. \quad (1)$$

PROOF. Let us define

$$\xi_1 = \mu_0(n_1) - N + n_1,$$

$$\xi_l = \mu_0(n_1 + \ldots + n_l) - \mu_0(n_1 + \ldots + n_{l-1}) + n_l, \quad l = 2, \ldots, t. \quad (2)$$

Introduce the generating function

$$\sum_{k_1, \ldots, k_t} \mathbf{P}\{\xi_1 = k_1, \ldots, \xi_t = k_t\} x_1^{k_1} \ldots x_t^{k_t} = A_{n,N}(x). \quad (3)$$

Starting from the definition of $A_{n,N}(x)$ and (2), we have

$$A_{n,N}(x) = \mathbf{M} x_1^{\xi_1} \ldots x_t^{\xi_t} = \\ = \mathbf{M} x_1^{\mu_0(n_1) - N + n_1} \prod_{l=2}^{t} x_l^{\mu_0(n_1 + \ldots + n_l) - \mu_0(n_1 + \ldots + n_{l-1}) + n_l} = \\ = x_1^{-N} x_1^{n_1} x_2^{n_2} \ldots x_t^{n_t} F_{n_1 \ldots n_t}\left(\frac{x_1}{x_2}, \frac{x_2}{x_3}, \ldots, \frac{x_{t-1}}{x_t}, x_t\right),$$

from which it follows that

$$\sum_{n_1, \ldots, n_t} A_{n,N}(x) \prod_{l=1}^{t} \frac{(N z_l)^{n_l}}{n_l!} = \\ = x_1^{-N} \Phi^N\left(x_1 z_1, \ldots, x_t z_t; \; \frac{x_1}{x_2}, \frac{x_2}{x_3}, \ldots, \frac{x_{t-1}}{x_t}, x_t\right). \quad (4)$$

Denoting by $A^N(z;\ x)$ the right-hand side of (4), we have by virtue of (2.3)

$$A(z; x) = 1 + \sum_{l=1}^{t} \frac{e^{z_l x_l} - 1}{x_l} \prod_{i=l+1}^{t} e^{z_i x_i}. \tag{5}$$

It follows from (4) and (5) that $A_{n,N}(x)$ can be written as the integral

$$A_{n,N}(x) = \frac{n_1! \ldots n_t!}{(2\pi i)^t N^{n_1+\ldots+n_t}} \oint_{C_t} \ldots \oint \frac{[A(z; x)]^N}{z^{n_1+1} \ldots z_t^{n_t+1}} dz_1 \ldots dz_t, \tag{6}$$

in which the integration is carried out over the circles C_t: $|z_l| = r_l$ for $l = 1,..., t$. To prove (1) it suffices to prove the convergence of multivariate generating functions; i.e., we need to show that for $0 \leqslant x_l \leqslant 1$, where $l = 1,..., t$,

$$A_{n,N}(x) \to \prod_{l=1}^{t} e^{(\theta_l - \theta_{l-1})(x_l - 1)}. \tag{7}$$

In investigating the asymptotic behavior of the integral in (6), we shall proceed as before, using the saddle-point method. Let $\frac{n_l^2}{2N} = \omega_l^2$, $\varepsilon = 1/\sqrt{N}$ and

$$f(z; x) = \ln A(z; x) - \sqrt{2}\,\varepsilon \sum_{l=1}^{t} \omega_l \ln z_l. \tag{8}$$

Then (6) can be written

$$A_{n,N}(x) = \frac{n_1! \ldots n_t!}{(2\pi i)^t N^{n_1+\ldots+n_t}} \oint_{C_t} \ldots \oint \frac{e^{Nf(z; x)}}{z_1 \ldots z_t} dz_1 \ldots dz_t. \tag{9}$$

Let $\overset{*}{z_l}$, $l = 1, \ldots, t$, denote a positive number satisfying the system of equations

$$\frac{\partial f}{\partial z_l} = \frac{1}{A}\cdot\frac{\partial A}{\partial z_l} - \frac{\sqrt{2}\,\varepsilon\omega_l}{z_l} = 0, \quad l = 1, \ldots, t. \tag{10}$$

(Such a solution exists: if $f(z;\ x)$ is regarded as a function of real positive variables $z_1,\ldots, z_t$, it has a minimum at the point $z_l = \overset{*}{z}_l$ for $l=1,\ldots, t$.) As $\varepsilon \to 0$, the solutions given by (10) tend to zero, so that we have the expansions in powers of ε:

$$\begin{gathered}\overset{*}{z}_l = z_{l1}\varepsilon + z_{l2}\varepsilon^2 + \ldots, \\ A^* = A(z^*;x) = A_0 + A_1\varepsilon + A_2\varepsilon^2 + \ldots, \\ \frac{\partial A^*}{\partial z_l} = A_{l0} + A_{l1}\varepsilon + \ldots .\end{gathered} \tag{11}$$

Here and below, expressions of the type $\frac{\partial A^*}{\partial z_l}$ are to be understood as short for $\left.\frac{\partial A(z;x)}{\partial z_l}\right|_{z=z^*}$. Let us substitute $\overset{*}{z}_l$ into the function (5) and let us express in terms of z_{lk} the first terms of the expansions of $A(z^*;\ x)$ and $\partial A^*/\partial z_l$ in powers of ε:

$$\begin{gathered}A_0 = 1, \quad A_1 = \sum_{l=1}^{t} z_{l1}, \\ A_2 \doteq \sum_{l=1}^{t} z_{l2} + \frac{1}{2}\sum_{l=1}^{t} x_l z_{l1}^2 + \sum_{l=1}^{t-1} z_{l1} \sum_{s=l+1}^{t} x_s z_{s1}, \\ A_{l0} = 1, \quad A_{l1} = \sum_{s=l}^{t} z_{s1}x_s + x_l \sum_{s=1}^{l-1} z_{s1}.\end{gathered} \tag{12}$$

Substituting (11) and (12) into (10), we find

$$\begin{gathered}z_{l1} = \sqrt{2}\,\omega_l, \\ z_{l2} = 2\omega_l\,[\omega_1 + \ldots + \omega_t - (\omega_l x_l + \ldots + \omega_t x_t) - \\ - x_l(\omega_1 + \ldots + \omega_{l-1})].\end{gathered} \tag{13}$$

Integrating in (9) over the contours $|z_l| = \overset{*}{z}_l$ and making the substitution $z_l = \overset{*}{z}_l e^{i\varphi_l}$, we can write (9) as

$$A_{n,N}(x) = \frac{n_1! \ldots n_t!}{(2\pi)^t N^{n_1+\ldots+n_t}} \times$$

$$\times \int_{-\pi}^{\pi} \ldots \int_{-\pi}^{\pi} \exp\left[Nf\left(z_1^* e^{i\varphi_1}, \ldots, z_t^* e^{i\varphi_t}; x\right)\right] d\varphi_1 \ldots d\varphi_t. \quad (14)$$

Denote by K the coefficient of the integral. Using Stirling's formula, we obtain

$$\ln K = -t \ln \sqrt{2\pi} + \ln\left(2^{t/4} \varepsilon^{t/2} N^{t/2} \sqrt{\omega_1 \ldots \omega_t}\right) +$$

$$+ \sum_{l=1}^{t} \sqrt{2}\, \omega_l \varepsilon N \ln\left(\sqrt{2}\, \omega_l \varepsilon\right) - \sqrt{2}\, \varepsilon N \sum_{l=1}^{t} \omega_l + O(\varepsilon). \quad (15)$$

Let us define $S = \{|\varphi_l| \leqslant \delta,\ l = 1, \ldots, t\}$, where δ is a positive constant, whose value will be chosen later, and $\overline{S} = \{|\varphi_l| \leqslant \pi,\ l = 1, \ldots, t\} \setminus S$. We express the integral in (14) as the sum

$$K \int_S \ldots \int e^{Nf} d\varphi_1 \ldots d\varphi_t + K \int_{\overline{S}} \ldots \int e^{Nf} d\varphi_1 \ldots d\varphi_t = I_1 + I_2.$$

In the domain $\overline{S}$, we have the inequality

$$\operatorname{Re} f\left(z_1^* e^{i\varphi_1}, \ldots, z_t^* e^{i\varphi_t}; x\right) \leqslant f\left(z_1^*, \ldots, z_t^*; x\right) - \Delta\varepsilon, \quad (16)$$

where $\Delta > 0$. In fact, by virtue of (5),

$$A\left(z_1^* e^{i\varphi_1}, \ldots, z_t^* e^{i\varphi_t}; x\right) / A\left(z_1^*, \ldots, z_t^*; x\right),$$

regarded as a function of $\varphi_1, \ldots, \varphi_t$, is the characteristic function of a t-dimensional lattice distribution concentrated on the lattice of points of t-dimensional space with integer coordinates. It can easily be seen from formula (5) that this characteristic function involves the terms

$$p_0 + p_1 e^{i\varphi_1} + \ldots + p_t e^{i\varphi_t}$$

with positive p_k (p_0 being a positive constant bounded from below) and $p_k \geqslant C\varepsilon > 0$, for $k=1,\ldots,t$, for sufficiently large N. For $0<\delta|\varphi_k|\leqslant\pi$, we have the inequality

$$|p_0 + p_k e^{i\varphi_k}| = |p_0 + p_k| \cdot \sqrt{1 - \frac{2p_0 p_k(1-\cos\varphi_k)}{(p_0+p_k)^2}} \leqslant$$
$$\leqslant (p_0 + p_k)(1 - \Delta_1\varepsilon), \qquad \Delta_1 > 0,$$

from which (16) follows since the functions f and A are related by (8). From (16) it follows that, as $N \to \infty$,

$$I_2 = K \exp\left[Nf\left(\overset{*}{z_1}, \ldots, \overset{*}{z_t}; x\right)\right] \cdot O\left(e^{-\sqrt{N}\Delta}\right).$$

We expand the function f in the exponent in the integrand of I_1 in a Taylor series about the point $z^* = \left(\overset{*}{z_1}, \ldots, \overset{*}{z_t}\right)$:

$$f(z; x) = f_0 + \sum_{l=1}^{t} f'_l\left(z_l - \overset{*}{z_l}\right) +$$
$$+ \frac{1}{2} \sum_{l,m=1}^{t} f''_{lm}\left(z_l - \overset{*}{z_l}\right)\left(z_m - \overset{*}{z_m}\right) + R, \quad (17)$$

where

$$f_0 = f\left(\overset{*}{z_1}, \ldots, \overset{*}{z_t}; x\right) = \ln A^* - \sqrt{2}\,\varepsilon \sum_{l=1}^{t} \omega_l \ln \overset{*}{z_l},$$
$$f'_l = \frac{1}{A^*} \cdot \frac{\partial A^*}{\partial z_l} - \frac{\sqrt{2}\,\varepsilon\omega_l}{\overset{*}{z_l}},$$
$$f''_{ll} = \frac{1}{A^*} \cdot \frac{\partial^2 A^*}{\partial z_l^2} - \left(\frac{1}{A^*}\frac{\partial A^*}{\partial z_l}\right)^2 + \frac{\sqrt{2}\,\varepsilon\omega_l}{\left(\overset{*}{z_l}\right)^2}, \quad (18)$$
$$f''_{lm} = \frac{1}{A^*}\frac{\partial^2 A^*}{\partial z_l \partial z_m} - \frac{1}{A^*} \cdot \frac{\partial A^*}{\partial z_l} \cdot \frac{\partial A^*}{\partial z_m}, \quad l \neq m.$$

The values $\overset{*}{z_l}$ satisfy equations (10); hence $f'_l = 0$. Since, for $l \geqslant m$, we have

$$\frac{\partial^2 A}{\partial z_l \partial z_m} = x_l \frac{\partial A}{\partial z_m} \quad \text{and} \quad \frac{1}{A^*} \frac{\partial A^*}{\partial z_l} = \frac{\sqrt{2}\,\varepsilon\omega_l}{z_l^*},$$

it follows that

$$\begin{aligned} f''_{ll} &= x_l \frac{\sqrt{2}\,\varepsilon\omega_l}{z_l^*} - \left(\frac{\sqrt{2}\,\varepsilon\omega_l}{z_l^*}\right)^2 + \frac{\sqrt{2}\,\varepsilon\omega_l}{(z_l^*)^2}, \\ f''_{lm} &= x_l \frac{\sqrt{2}\,\varepsilon\omega_m}{z_m^*} - 2\varepsilon^2 \frac{\omega_l}{z_l^*} \cdot \frac{\omega_m}{z_m^*}, \quad l > m. \end{aligned} \tag{19}$$

The remainder term R in (17) can be expressed as

$$R = O(\varepsilon^3) + O\left(\varepsilon \sum_{l=1}^{t} |\varphi_l|^3\right). \tag{20}$$

In fact, by virtue of (8) we can decompose f as the sum of several terms. The first term $\ln A$ is an analytic function of $z_1, \ldots, z_t$ in a neighborhood of the origin. Hence, keeping terms through second-degree in the expansion of this term $\ln A$, we obtain $O(\varepsilon^3)$ as the remainder. The rest of the remainder term in (20) can be handled as follows. Let $z = re^{i\varphi}$. Then in the expansion

$$\ln z = \ln r + \frac{z-r}{r} - \frac{1}{2}\frac{(z-r)^2}{r^2} + R_1,$$

the remainder term R_1 is equal to

$$R_1 = 1 + i\varphi - e^{i\varphi} + (e^{i\varphi} - 1)^2/2,$$

from which it follows that $R_1 = O(|\varphi|^3)$.

Substituting (11) and (12) into (18) and (19), and expanding f_0 in a series in ε, we obtain

$$f_0 = \ln(1 + A_1\varepsilon + A_2\varepsilon^2 + \ldots) - \sqrt{2}\varepsilon \sum_{l=1}^{t} \omega_l \ln(\sqrt{2}\omega_l\varepsilon) -$$

$$-\sqrt{2}\varepsilon \sum_{l=1}^{t} \omega_l \ln\left(1 + \frac{z_{l2}\varepsilon}{z_{l1}} + \ldots\right) = A_1\varepsilon + \varepsilon^2\left(A_2 - \frac{A_1^2}{2}\right) -$$

$$-\sqrt{2}\varepsilon \sum_{l=1}^{t} \omega_l \ln(\sqrt{2}\omega_l\varepsilon) - \varepsilon^2 \sum_{l=1}^{t} z_{l2} + O(\varepsilon^3), \quad (21)$$

$$f''_{ll} = \frac{1}{\sqrt{2}\varepsilon\omega_l} + O(1), \quad f''_{lm} = O(1), \quad l \neq m.$$

From (15) and (21) we have

$$\ln K + Nf_0 = \left(A_2 - \frac{A_1^2}{2} - \sum_{l=1}^{t} z_{l2}\right) +$$

$$+ \ln\left(N^{t/4} 2^{t/4} \sqrt{\omega_1 \ldots \omega_t}\right) - t \ln\sqrt{2\pi} + O(\varepsilon) =$$

$$= \sum_{l=1}^{t} (\theta_l - \theta_{l-1})(x_l - 1) + \ln\left(N^{t/4} 2^{t/4} \sqrt{\omega_1 \ldots \omega_t}\right) -$$

$$- t \ln\sqrt{2\pi} + O(\varepsilon) + o(1), \quad (22)$$

from which it follows, in particular, that

$$I_2 = O(N^{t/4} e^{-\Delta\sqrt{N}}).$$

Let us represent the integral I_1 as the product:

$$I_1 = I \cdot \exp\left[\sum_{l=1}^{t} (\theta_l - \theta_{l-1})(x_l - 1)\right],$$

where

$$I = \frac{N^{t/4} \cdot 2^{t/4} \sqrt{\omega_1 \ldots \omega_t}}{(2\pi)^{t/2}} \times$$

$$\times \int_{|\varphi_l| \leqslant \delta} \!\!\ldots\! \int \exp\left\{-\frac{N}{2}\left[\sum_{l,m=1}^{t} f''_{lm}(z_l - z_l^*)(z_m - z_m^*) + \right.\right.$$

$$\left.\left. + O(\varepsilon^3) + O\left(\varepsilon \sum_{l=1}^{t} |\varphi_l|^3\right)\right]\right\} d\varphi_1 \ldots d\varphi_t. \quad (23)$$

Making the change of variables $y_l = (2N)^{1/4}\sqrt{\omega_l}\cdot\varphi_l$, in (23), we have, as $N \to \infty$,

$$I = \left(\frac{1}{\sqrt{2\pi}}\right)^t \int_{|y| \leqslant \delta(2N)^{1/4}\sqrt{\omega_l}} \cdots \int \exp\left[-\frac{1}{2}\sum_{l=1}^{t} y_l^2 + \right.$$

$$\left. + O\left(\sum_l \frac{|y_l|^3}{\sqrt[4]{N}}\right) + O(\varepsilon)\right] dy_1 \ldots dy_t. \quad (24)$$

Let us partition the integration domain $S = \{|y_l| \leqslant \delta(2N)^{1/4}\sqrt{\omega_l}, l = 1, \ldots, t\}$ into two parts: $S_0 = \{|y_l| \leqslant \delta N^{1/16}, l = 1, \ldots, t\}$ and $S \setminus S_0$. It is easy to see that in integrating over S_0 we obtain unity in the limit as $N \to \infty$; integrating over $S \setminus S_0$ we see that the limit can be made arbitrarily small by choosing δ small enough. The theorem is proved.

§4. Convergence to a Gaussian Process in the Central Domain

Let us consider n, $N \to \infty$ in the central domain. Introduce a new time $\theta = n/N$. For $t = 2$, the generating function (2.3) is of the following form:

$$\Phi_N(z; x) = [(e^{z_1} - 1)e^{z_2} + (e^{z_2} - 1)x_1 + x_1x_2]^N. \quad (1)$$

The mixed second derivative of this generating function with respect to x_1 and x_2 at the point $x_1 = x_2 = 1$ is equal to

$$Ne^{(z_1+z_2)(N-1)} + N(N-1)e^{z_1(N-2)} \cdot e^{z_2(N-1)}. \quad (2)$$

Let us expand (2) in powers z_1 and z_2; the coefficient of $z_1^{n_1}z_2^{n_2}$ multiplied by $n_1!\,n_2!/N^{n_1+n_2}$ yields

$$\mathbf{M}\mu_0(n_1)\mu_0(n_1+n_2) = N\left(1-\frac{1}{N}\right)^{n_1+n_2} + \\ + N(N-1)\left(1-\frac{2}{N}\right)^{n_1}\left(1-\frac{1}{N}\right)^{n_2}. \quad (3)$$

Let n_1, n_2, $N\to\infty$ in such a way that $\frac{n_1}{N}\to\theta$ and $\frac{n_1+n_2}{N}\to\theta'$, where $\theta<\theta'$. Then from (3) and the asymptotic behavior of $\mathbf{M}\mu_0$ in (1.1.14), we can obtain an asymptotic formula for the covariance (we assume from now on that $n=n_1$, $n'=n_1+n_2$):

$$\mathbf{Cov}\,(\mu_0(n), \mu_0(n')) \sim Ne^{-\theta'}(1-(1+\theta)e^{-\theta}), \quad \theta'>\theta, \quad (4)$$

and the limiting correlation coefficient

$$\rho\,(\mu_0(n), \mu_0(n')) \sim \rho\,(\theta, \theta') = \frac{e^{-(\theta'-\theta)}\sqrt{e^{-\theta}(1-(1+\theta)e^{-\theta})}}{\sqrt{e^{-\theta'}(1-(1+\theta')e^{-\theta'})}}, \\ \theta'\geqslant\theta,$$

which can be written as

$$\rho\,(\theta, \theta') = \frac{V(\theta)}{V(\theta')}, \qquad \theta'\geqslant\theta,$$

where the function

$$V(\theta) = \sqrt{e^{\theta}-1-\theta} \sim \sqrt{N}\,\frac{\sqrt{\mathbf{D}\mu_0(n)}}{\mathbf{M}\mu_0(n)}, \qquad \frac{n}{N}=\theta,$$

characterizes the limit behavior of $\mu_0(n)$.

We introduce now the normalized random process

$$\xi(\theta) = \frac{\mu_0(n)-\mathbf{M}\mu_0(n)}{\sqrt{N}}, \qquad \frac{n}{N}\leqslant\theta<\frac{n+1}{N}. \quad (5)$$

Theorem 1. *Finite-dimensional distributions of the process* $\xi(\theta)$ *converge, as* $n, N \to \infty$, *in the central domain to finite-dimensional distributions of a Gaussian process with correlation function*

$$e^{-\theta'}(1-(1+\theta)e^{-\theta}), \quad \theta' > \theta. \tag{6}$$

PROOF. Let $\theta_1^* < \ldots < \theta_t^*$, $n_1 + \cdots + n_l = [N\theta_l^*] = N\theta_l$. In the definition of $\xi(\theta)$ in (5) we replace $\mathbf{M}\mu_0(n)$ by its asymptotic form $Ne^{-\theta}$; since $\mathbf{M}\mu_0(n) - Ne^{-\theta} = O(1)$, the limit distribution will be the same. Assume next $\varepsilon = 1/\sqrt{N}$. The characteristic function of these modified random variables $\xi(\theta_1)$, $\xi(\theta_2), \ldots, \xi(\theta_t)$ (we do not introduce new notation for them) can be written with the help of (2.3) as

$$A_{n,N}(\tau_1, \ldots, \tau_t) = \mathbf{M}\exp\left[i \sum_{k=1}^{t} \tau_k \xi(\theta_k^*)\right] =$$

$$= \frac{\prod_{l=1}^{t} n_l! \exp\left\{-\frac{i}{\varepsilon} \sum_{l=1}^{t} \tau_l e^{-\theta_l^*}\right\}}{(2\pi i)^t N^{n_1+\ldots+n_t}} \times$$

$$\times \oint \ldots \oint_{C_t} \Phi^N\left(z; e^{i\tau_1\varepsilon}, \ldots, e^{i\tau_t\varepsilon}\right) \frac{dz_1 \ldots dz_t}{z_1^{n_1+1} \ldots z_t^{n_t+1}}. \tag{7}$$

We shall choose the integration contours C_t as follows: $z_l = \omega_l e^{i\varphi_l}$, where $\omega_l = \theta_l - \theta_{l-1}$ for $l = 1, \ldots, t$, where $\theta_0 = 0$. Let us take

$$f = \ln \Phi\left(\omega_1 e^{i\varphi_1}, \ldots, \omega_t e^{i\varphi_t}; e^{i\tau_1\varepsilon}, \ldots, e^{i\tau_t\varepsilon}\right) -$$

$$- \sum_{l=1}^{t} \omega_l \ln \omega_l e^{i\varphi_l} - i\varepsilon \sum_{l=1}^{t} \tau_l e^{-\theta_l^*}. \tag{8}$$

We divide into two parts the integral obtained from (7) after replacing z_l by $\omega_l e^{i\varphi_l}$:

$$A_{n,N}(\tau_1, \ldots, \tau_t) = \frac{\prod_{l=1}^{t} n_l!}{(2\pi)^t N^{n_1+\ldots+n_t}} \times$$

$$\times \left[\int_S \cdots \int e^{Nf} d\varphi_1 \ldots d\varphi_t + \int_{\bar{S}} \cdots \int e^{Nf} d\varphi_1 \ldots d\varphi_t \right], \tag{9}$$

where $S = \{\varphi_l : |\varphi_l| \leqslant \delta,\ l = 1, \ldots, t\}$, and $\bar{S} = \{|\varphi_l| \leqslant \pi,\ l = 1, \ldots, t\} \setminus S$.

In the same way as Theorem 3.1, we can show that

$$\int_{\bar{S}} \cdots \int e^{Nf} d\varphi_1 \ldots d\varphi_t = \exp[Nf(\omega_1, \ldots, \omega_t;\ 1, \ldots, 1)] \cdot O(q^N), \tag{10}$$

where $0 < q < 1$. The factor $K = \frac{1}{(2\pi)^t N^{n_1+\ldots+n_t}} \prod_{l=1}^{t} n_l!$ in front of the integral in (9) is estimated by Stirling's formula:

$$\ln K = N \sum_{l=1}^{t} \omega_l (\ln \omega_l - 1) + \frac{1}{2} \ln \left(N^t \prod_{l=1}^{t} \omega_l \right) - - t \ln \sqrt{2\pi} + o(1). \tag{11}$$

Let us expand the function f in the domain S in powers of φ_l and ε, taking

$$z_l = \omega_l e^{i\varphi_l} = \omega_l \left(1 + i\varphi_l - \frac{\varphi_l^2}{2} + \ldots \right)$$

and

$$e^{i\tau_l \varepsilon} = 1 + i\tau_l \varepsilon - \frac{\tau_l^2 \varepsilon^2}{2} + \ldots :$$

$$f=\theta_t-\sum_{l=1}^{t}\omega_l\ln\omega_l-\sum_{l=1}^{t}\omega_l\frac{\varphi_l^2}{2}+\varepsilon\sum_{l=1}^{t}\tau_l\sum_{k=1}^{l}\omega_k\varphi_k e^{-\theta_k}-$$
$$-\frac{\varepsilon^2}{2}\left[\sum_{l=1}^{t}\left(e^{-\theta_l}-e^{-\theta_{l+1}}\right)(\tau_1+\ldots+\tau_l)^2-\left(\sum_{l=1}^{t}\tau_l e^{-\theta_l}\right)^2\right]+$$
$$+O(\varepsilon^3)+O\left(\sum_{l=1}^{t}|\varphi_l|^3\right),\quad \theta_{t+1}=\infty. \quad (12)$$

Letting $\sqrt{N}\varphi_l=y_l$, for $l=1, \ldots, t$, and substituting (11) and (12) in the first integral in (9), we get

$$K\int_S\ldots\int e^{Nf}d\varphi_1\ldots d\varphi_t=\frac{e^{-Q/2}\sqrt{\omega_1\ldots\omega_t}}{(2\pi)^{t/2}}\times$$
$$\times\int_{|y_l|\leqslant\delta\sqrt{N}}\ldots\int\exp\left[-\frac{1}{2}\sum_{l=1}^{t}\omega_l\left(y_l+\sum_{k=l}^{t}\tau_k e^{-\theta_k}\right)^2+\right.$$
$$\left.+O\left(\sum_{l=1}^{t}|y_l|^3\varepsilon\right)+O(\varepsilon)+o(1)\right]dy_1\ldots dy_t, \quad (13)$$

where

$$Q=\sum_{l=1}^{t}\left(e^{-\theta_l}-e^{-\theta_{l+1}}\right)(\tau_1+\ldots+\tau_l)^2-\left(\sum_{l=1}^{t}\tau_l e^{-\theta_l}\right)^2-$$
$$-\sum_{l=1}^{t}\omega_l\left(\sum_{k=l}^{t}\tau_k e^{-\theta_k}\right)^2,\quad \theta_{t+1}=\infty. \quad (14)$$

Breaking up the integration domain in (13) into the domain $\{|y_l|\leqslant\delta N^{1/8},\ l=1, \ldots, t\}$ and the complementary domain, we have, as with the estimate (3.24), the result that, for sufficiently small δ, the expression (13) converges, as $N\longrightarrow\infty$, to the function $e^{-Q/2}$. From (10), (11), and (12) it follows that, for any $\delta>0$,

$$K \int \underset{\overline{S}}{\cdots} \int e^{Nf} \, d\varphi_1 \ldots d\varphi_t = O\left(N^{t/2} q^t\right).$$

Hence, the characteristic function (9) converges to $e^{-Q/2}$, which is then the characteristic function of the limiting multidimensional normal distribution with the correlation function (6). The theorem is proved

Suppose now that the random process $\xi(\theta)$ is defined by (5) only at points of the form $\theta = n/N$. Then we interpolate this process linearly over the whole interval. We denote by $\eta(\theta)$ the Gaussian process with correlation function given by (6).

Theorem 2. *If $T_2 > T_1 > 0$ and if g is a continuous functional on the space of continuous functions $C[T_1, T_2]$, then under the conditions of Theorem 1, the distribution $g(\xi(\cdot))$ converges to the distribution $g(\eta(\cdot))$, where $\eta(\theta)$ is the Gaussian process defined above.*

PROOF. We need only to prove that

$$\mathbf{M}(\xi(\theta_1) - \xi(\theta_2))^4 \leqslant C(\theta_1 - \theta_2)^2, \tag{15}$$

where C is some constant (see, for example, I. I. Gikhman and A. V. Skorokhod [14], Chapter 9, Section 2, Theorem 2). Assuming in (1) that $x_1 = x$, $x_2 = 1/x$, we get the generating function for $\zeta = \mu_0(n_1) - \mu_0(n_1 + n_2)$:

$$\sum_{n_1, n_2} \frac{N^{n_1+n_2} z_1^{n_1} z_2^{n_2}}{n_1! \, n_2!} \sum_k \mathbf{P}\{\zeta = k\} x^k =$$
$$= [(e^{z_1} - 1) e^{z_2} + (e^{z_2} - 1) x + 1]^N. \tag{16}$$

Differentiating this generating function s times with respect to x and setting $x = 1$, we obtain

$$N^{[s]} e^{(z_1+z_2)(N-s)} (e^{z_2} - 1)^s.$$

Expanding this expression in powers of z_1 and z_2, we find, from the definition (16), the factorial moments

$$\mathbf{M}\zeta^{[s]} = N^{[s]}\left(\frac{N-s}{N}\right)^{n_1}\sum_{l=0}^{s}(-1)^l C_s^l\left(1-\frac{l}{N}\right)^{n_2}. \qquad (17)$$

Since, for $n_1/N=\theta_1$ and $(n_1+n_2)/N=\theta_2$,

$$\xi(\theta_1)-\xi(\theta_2)=\frac{\zeta-\mathbf{M}\zeta}{\sqrt{N}},$$

we have

$$\mathbf{M}(\xi(\theta_2)-\xi(\theta_1))^4=\frac{\mathbf{M}(\zeta-\mathbf{M}\zeta)^4}{N^2}.$$

From this, using (17), we can obtain via cumbersome but rather standard calculations the estimate

$$\mathbf{M}(\xi(\theta_1)-\xi(\theta_2))^4=O((\theta_1-\theta_2)^2)+O\left(\frac{\theta_2-\theta_1}{N}\right). \qquad (18)$$

Since for $n_2 \geqslant 1$ we have $n_2/N=\theta_2-\theta_1 \geqslant 1/N$, inequality (15) follows from (18). The estimate (15) is trivial in the case where θ_1 and θ_2 are taken from one interpolation interval or neighboring intervals.

Remark. If instead of the normalization (5) we take the normalization

$$\xi_1(\theta)=\frac{\mu_0(n)-\mathbf{M}\mu_0(n)}{\sqrt{\mathbf{D}\mu_0(n)}}$$

and if we make the change of time $t=\ln V(\theta)$ where $\theta=n/N$ and $V(\theta)=\sqrt{e^\theta-1-\theta}$, the stationary Gaussian process $\eta_1(t)$ with correlation function $\mathbf{M}\eta(t_1)\eta(t_2)=e^{-|t_1-t_2|}$ will be the limit process for $\xi_1(\theta(t))$.

§5. Convergence to a Poisson Process in the Right-Hand Domain

In this section, we shall examine n, $N \longrightarrow \infty$ for which the difference $\frac{n}{N} - \ln N$ is bounded. Introduce a new time $\theta = Ne^{-n/N}$. Note that this new time is "reciprocal," i.e., the larger n, the smaller θ. Let us define θ_l by

$$\frac{n_1 + \ldots + n_l}{N} - \ln N = -\ln \theta_l\,, \qquad l = 1, \ldots, t. \tag{1}$$

Theorem 1. *If* n_l, $N \longrightarrow \infty$ *as* $N \longrightarrow \infty$, *for* $l=1, \ldots, t$, *in such a way that the variables* θ_l *defined by* (1) *are bounded, the random variables*

$$\mu_0(n_1 + \ldots + n_t),\ \mu_0(n_1 + \ldots + n_l) - \mu_0(n_1 + \ldots + n_{l+1}),$$
$$l=1, \ldots, t-1,$$

are asymptotically independent and their distributions converge to Poisson distributions with parameters θ_t *and* $\theta_l - \theta_{l+1}$ *for* $l=1, \ldots, t-1$, *respectively.*

PROOF. Let $\xi_t = \mu_0(n_1 + \ldots + n_t)$ and let

$$\xi_l = \mu_0(n_1 + \ldots + n_l) - \mu_0(n_1 + \ldots + n_{l+1}),$$

for $l=1, \ldots, t-1$. Replacing x_l in (2.3) with x_l/x_{l-1} for $l=2, \ldots, t$, we easily obtain the generating function

$$\sum_{n_1, \ldots, n_t} \sum_{k_1, \ldots, k_t} \prod_{l=1}^{t} \left[\frac{(z_l N)^{n_l}}{n_l!} x_l^{k_l} \right] \mathbf{P}\{\xi_i = k_i\,,\ i = 1, \ldots, t\} =$$
$$= \left[x_t + \sum_{l=1}^{t} \left(e^{z_l} - 1\right) e^{z_{l+1} + \ldots + z_t} x_{l-1} \right]^N, \qquad x_0 = 1. \tag{2}$$

We shall prove the theorem by the method of moments. Differentiating (2) s_l times with respect to x_l, for $l=1, \ldots, t$, we have at the point $x_1= \ldots =x_t=1$

$$N^{[s]}\, e^{z_1(N-s)} \prod_{l=1}^{t-1} e^{z_{l+1}(N-s+s_1+\ldots+s_{l-1})} \left(e^{z_{l+1}}-1\right)^{s_l}, \quad (3)$$

where $s_0=0$ and $s=s_1+ \ldots +s_t$. From (2) and (3), we deduce the expression for factorial moments

$$\mathbf{M}\xi_1^{[s_1]} \ldots \xi_t^{[s_t]} =$$

$$= N^{[s]}\left(1-\frac{s}{N}\right)^{n_1} \prod_{l=1}^{t-1}\left[\sum_{r_l=0}^{s_l} C_{s_l}^{r_l}(-1)^{s_l-r_l} \times \right.$$

$$\left. \times \left(1-\frac{s-s_1-\ldots-s_{l-1}-r_l}{N}\right)^{n_{l+1}}\right]. \quad (4)$$

It follows from the definition (1) that $n_1/N=\ln N-\ln \theta_1$ and $n_{l+1}/N=\ln(\theta_l/\theta_{l+1})$, Hence, as $n, N\to\infty$, we have $N^{[s]}\left(1-\frac{s}{N}\right)^{n_1}\to \theta_1^s$ and

$$\sum_{r=0}^{s_l} C_{s_l}^{r}(-1)^{s_l-r}\left(1-\frac{s-s_1-\ldots-s_{l-1}-r}{N}\right)^{n_{l+1}} \to$$

$$\to \sum_{r=0}^{s_l} C_{s_l}^{r}(-1)^{s_l-r} \exp\left[-(s-s_1-\ldots-s_{l-1}-r)\ln\frac{\theta_l}{\theta_{l+1}}\right]=$$

$$=\left(\frac{\theta_{l+1}}{\theta_l}\right)^{s-s_1-\ldots-s_{l-1}}\left(\frac{\theta_l}{\theta_{l+1}}-1\right)^{s_l}=\frac{\theta_{l+1}^{s-s_1-\ldots-s_l}\,(\theta_l-\theta_{l+1})^{s_l}}{\theta_l^{s-s_1-\ldots-s_{l-1}}}.$$

Therefore, the limit given by (4) is equal to

$$\lim \mathbf{M}\xi_1^{[s_1]} \ldots \xi_t^{[s_t]} = \theta_t^{s_t}\prod_{l=1}^{t-1}(\theta_l-\theta_{l+1})^{s_l}. \quad (5)$$

Since the right-hand side of (5) is equal to the product of the factorial moments of Poisson independent random variables, the theorem is proved.

§6. Convergence to Gaussian Processes in Intermediate Domains

We consider first the right-hand intermediate domain, i.e., $n,\ N \to \infty$ such that $\frac{n}{N} \to \infty,\ \frac{n}{N} - \ln N \to -\infty$. It is not feasible to introduce a new time scale for the entire right-hand intermediate domain and hence we shall introduce "local times." Let the parameter $\alpha_0 = n_0/N$ vary in such a way that $\alpha_0 \to \infty$ and $\alpha_0 - \ln N \to -\infty$. We examine $n \to \infty$ such that $\theta = \frac{n}{N} - \alpha_0$ is bounded. From Theorem 1.1.1 we have the asymptotic formula

$$\mathbf{M}\mu_0(n) \sim Ne^{-(\alpha_0+\theta)}. \tag{1}$$

It can easily be seen that the asymptotic formula (4.4) holds for covariances in the right-hand intermediate domain as well; that is, in the notation introduced above,

$$\mathbf{Cov}\,(\mu_0(n), \mu_0(n')) \sim Ne^{-\alpha_0-\theta'} \tag{2}$$

if $\theta' = \frac{n'}{N} - \alpha_0 \geqslant \theta = \frac{n}{N} - \alpha_0$. From this we obtain the limiting correlation coefficient

$$\rho\,(\mu_0(n), \mu_0(n')) \sim e^{-\frac{1}{2}|\theta'-\theta|}.$$

Let us introduce the normalized random process

$$\xi(\theta) = \frac{\mu_0(n) - \mathbf{M}\mu_0(n)}{\sqrt{Ne^{-\alpha_0}}}, \qquad \frac{n}{N} \leqslant \theta < \frac{n+1}{N}. \tag{3}$$

Theorem 1. *If* $\alpha_0 \to \infty$ *in such a way that* $\alpha_0 - \ln N \to -\infty$ *and* $\frac{n}{N} - \alpha_0 = \theta$ *is bounded, the finite-dimensional distributions of the process* $\xi(\theta)$ *converge to finite-dimensional distributions of a Gaussian process* $\eta(\theta)$ *with correlation function* $\mathbf{M}\eta(\theta)\eta(\theta') = e^{-\theta'}$, *where* $\theta' \geqslant \theta$.

PROOF. Let $\overset{*}{\theta}_1 < \ldots < \overset{*}{\theta}_t$, $n_1 + \cdots + n_k = [N(\overset{*}{\theta}_k + \alpha_0)] = N(\theta_k + \alpha_0)$. The characteristic function

$$A_{n,N}(\tau_1, \ldots, \tau_t) = \mathbf{M}\exp\left\{i \sum_{k=1}^{t} \tau_k \xi(\overset{*}{\theta}_k)\right\} \tag{4}$$

is representable in the form of the integral in (4.7) if we take $\varepsilon = (Ne^{-\alpha_0})^{-1/2}$. Further, we find the limit of the integral in (4.7) using a method similar to that applied in Section 4. Here we note briefly the steps in the proof which are different from those given in Section 4. Let us replace the variables z_k by $\omega_k e^{i\varphi_k}$, where $\omega_1 = \alpha_0 + \theta_1$ and $\omega_k = \theta_k - \theta_{k-1}$, for $k = 2, 3, \ldots, t$. Then

$$A_{n,N}(\tau_1, \tau_2, \ldots, \tau_t) = K \int_{-\pi}^{\pi} \ldots \int_{-\pi}^{\pi} e^{Nf} d\varphi_1 \ldots d\varphi_t, \tag{5}$$

where

$$f = \ln \Phi(\omega_1 e^{i\varphi_1}, \ldots, \omega_t e^{i\varphi_t};\ e^{i\tau_1\varepsilon}, \ldots, e^{i\tau_t\varepsilon}) -$$
$$- \frac{1}{N}\sum_{k=1}^{t} i\tau_k \varepsilon^{-1} e^{-\overset{*}{\theta}_k} - \sum_{k=1}^{t} \omega_k \ln(\omega_k e^{i\varphi_k});$$

the constant K is the same as in Section 4. The asymptotic behavior of $\ln K$ is also given by (4.11). As in (4.9), we express the integral as the sum of the integrals over the two subdomains S and S. In the domain S, let us expand the function f in powers ε and φ_k:

$$f = \sum_{k=1}^{t} \Big\{ \omega_k (1 - \ln \omega_k) - \frac{1}{2} \omega_k \varphi_k^2 + \omega_k \varphi_k \varepsilon e^{-\alpha_0} \sum_{m=k}^{t} \tau_m e^{-\theta_m} -$$
$$- \frac{1}{2} \varepsilon^2 e^{-\alpha_0 - \theta_k} [(\tau_1 + \ldots + \tau_k)^2 - (\tau_1 + \ldots + \tau_{k-1})^2] \Big\} +$$
$$+ O\left(\sum_{k=1}^{t} \omega_k |\varphi_k|^3 \right) + o\left(\frac{1}{N} \right).$$

Next we replace the variables $\varphi_k \sqrt{N\omega_k} = y_k$ for $k = 1, 2, \ldots, t$ and express the integral over S in the following form:

$$K \int \underset{S}{\ldots} \int e^{Nf} d\varphi_1 \ldots d\varphi_t = e^{-Q/2} I, \tag{6}$$

where

$$Q = \sum_{k=1}^{t} e^{-\theta_k} [(\tau_1 + \ldots + \tau_k)^2 - (\tau_1 + \ldots + \tau_{k-1})^2] =$$
$$= \sum_{k=1}^{t} \tau_k^2 e^{-\theta_k} + 2 \sum_{k<m} \tau_k \tau_m e^{-\theta_m}, \tag{7}$$

$$I = \frac{1}{(2\pi)^t} \underset{\substack{|y_k| \leqslant \delta \sqrt{N\omega_k} \\ k=1,\ldots,t}}{\int \cdots \int} \exp \Big\{ -\frac{1}{2} \sum_{k=1}^{t} \Big(y_k - \sqrt{\omega_k e^{-\alpha_0}} \times$$
$$\times \sum_{m=k}^{t} \tau_m e^{-\theta_m} \Big)^2 + O\left(\sum_{k=1}^{t} \frac{|y_k|^3}{\sqrt{N\omega_k}} \right) + o(1) \Big\} dy_1 \ldots dy_t. \tag{8}$$

Dividing the integration domain in (8) into

$$S_0 = \{ y_k : |y_k| \leqslant \delta \sqrt[8]{N\omega_k}, \quad k = 1, 2, \ldots, t \}$$

and a complementary domain, we see that $I \to 1$ as $n, N \to \infty$ if δ is sufficiently small. As in Section 4, we can show that

$$K \int \underset{\overline{S}}{\ldots} \int e^{Nf} d\varphi_1 \ldots d\varphi_t \to 0.$$

Hence the characteristic function $A_{n,N}(\tau_1, \ldots, \tau_t)$ converges to $e^{Q/2}$, i.e., the characteristic function of a multidimensional normal distribution. The theorem is proved.

We note that the limiting Gaussian process $\eta(\theta)$ with correlation function

$$\mathbf{M}\eta(\theta)\eta(\theta') = e^{-\max(\theta, \theta')}$$

is a process with independent increments if we use "reciprocal" time. In fact, for $\theta' > \theta$ and

$$\mathbf{M}\eta(\theta')(\eta(\theta) - \eta(\theta')) = 0,$$

i.e., $\eta(\theta')$ and $\eta(\theta) - \eta(\theta')$ are independent, $\mathbf{D}\eta(\theta) = e^{-\theta}$, and $\mathbf{D}\{\eta(\theta) - \eta(\theta')\} = e^{-\theta} - e^{-\theta'}$, where $\theta' \geqslant \theta$.

The process $\xi(\theta)$ defined by (3) depends not only on the parameter θ but also on the way α_0 varies. If α_0 and α_0' vary in such a way that $\alpha_0, \alpha_0' \to \infty$ and $|\alpha_0 - \alpha_0'| \to \infty$, by similar arguments we can prove that the finite-dimensional distributions of the corresponding processes $\xi_{\alpha_0}(\theta)$ and $\xi_{\alpha_0'}(\theta)$ converge to finite-dimensional distributions of independent Gaussian processes. Therefore, each $\alpha_0 \longrightarrow \infty$ is associated with its own local time θ and the pertinent limiting Gaussian process $\eta(\theta)$.

A similar situation is observed in the left-hand intermediate domain. Let the parameter $\alpha_0 = n_0/N$ vary as n_0, $N \longrightarrow \infty$ in such a way that $\alpha_0 \to 0$ and $N\alpha_0^2 \to \infty$. We shall consider $n \to \infty$ such that $n/N = \alpha_0\theta$ and the variable $\ln\theta$ is bounded. In the new time θ, we examine the normalized process

$$\xi(\theta) = \frac{\mu_0(n) - \mathbf{M}\mu_0(n)}{\alpha_0\sqrt{N/2}}, \quad \frac{n}{N} \leqslant \theta < \frac{n+1}{N}. \tag{9}$$

Theorem 2. *If* $n_0, n, N \longrightarrow \infty$ *in such a way that* $\alpha_0 = \frac{n_0}{N} \to 0$, $N\alpha_0^2 \to \infty$, $n/N = \alpha_0\theta$, *and* $\ln\theta$ *is bounded, the finite-dimensional distributions of the process* $\xi(\theta)$ *defined by* (9) *converge to finite-dimensional distributions of a Gaussian process* $\eta(\theta)$ *with correlation function* $\mathbf{M}\eta(\theta)\eta(\theta')=\theta^2$, *where* $\theta'>\theta$.

Let us note that the limiting Gaussian process $\eta(\theta)$ is a process with independent increments and that $\mathbf{D}\eta(\theta)=\theta^2$.

The proof of this theorem is based on the following lemma:

Lemma 1. *If the random variables* $\eta_k(N)$, *for* $k=1, \ldots, t$, *are independent and are asymptotically Gaussian as* $N\longrightarrow\infty$ *with parameters* $(0, \sigma_k)$, *then the joint distribution of the random variables*

$$\xi_k(N)=\eta_1(N)+\eta_2(N)+\ldots+\eta_k(N),\ k=1, \ldots, t,$$

is asymptotically Gaussian as $N\longrightarrow\infty$ *with covariance matrix* $\|b_{km}\|$, *where* $b_{km}=\sigma_k^2$, $m\geqslant k$.

Proof of Theorem 2. As in Section 1.2, let ν_k denote the number of throws as a result of which exactly k cells are occupied by particles for the first time ($N-k$ cells remaining empty). Let us express ν_k as the sum of independent random variables $\nu_k=\delta_1+\delta_2+\ldots+\delta_k$, where $\delta_1=\nu_1=1$ and $\delta_m=\nu_m-\nu_{m-1}$ for $m=2, 3, \ldots, k$. We showed in Theorem 1.2.2 that, if $k/N\longrightarrow\gamma$, where $0\leqslant\gamma<1$, and $k^2/N\longrightarrow\infty$, we have the asymptotic formulas

$$\mathbf{M}\nu_k=N\ln\frac{N}{N-k}+O(1),\quad \mathbf{D}\nu_k=\frac{Nk}{N-k}-N\ln\frac{N}{N-k}+O(1),$$

$$\mathbf{M}|\delta_l-\mathbf{M}\delta_l|^3=O\left(\frac{l}{N}\right),\ l=1,\ldots,k.$$

Suppose that n_k, $N\longrightarrow\infty$, for $k=1, 2, \ldots, t$, in such a way that

$$\frac{n_1+n_2+\ldots+n_k}{N}=\alpha_0\theta_k,\qquad k=1, 2, \ldots, t, \tag{10}$$

where $\alpha_0\longrightarrow 0$ and $N\alpha_0^2\rightarrow\infty$, and the variables $\ln\theta_k$ are bounded. In this case,

$$\mathbf{M}\mu_0(n_1 + \ldots + n_k) \sim Ne^{-\alpha_0\theta_k},$$

$$\mathrm{Cov}\,(\mu_0(n_1 + \ldots + n_k), \mu_0(n_1 + \ldots + n_m)) \sim \frac{N}{2}\alpha_0^2\theta_k^2, \quad m \geqslant k.$$

(See Theorem 1.1.1 and the asymptotic formula given in (4.4).)

From the definitions of the variables μ_0 and ν_k we have:

$$\mathbf{P}\{\mu_0(n_1) \leqslant N - k_1, \ \mu_0(n_1+n_2) \leqslant N - k_2, \ldots, \mu_0(n_1 + \ldots + n_t) \leqslant N - k_t\} = \mathbf{P}\{\nu_{k_1} \leqslant n_1, \ \nu_{k_2} \leqslant n_1 + n_2, \ldots, \nu_{k_t} \leqslant n_1 + n_2 + \ldots + n_t\}. \tag{11}$$

Put

$$n_1 + n_2 + \ldots + n_m = N \ln\frac{N}{N - k_m} + x_m\sqrt{\frac{Nk_m}{N - k_m} - N\ln\frac{N}{N - k_m}} + O(1), \quad m = 1, \ldots, t. \tag{12}$$

Solving (12) for k_m, we get

$$k_m = N(1 - e^{-\alpha_0\theta_m}) - x_m\alpha_0\theta_m\sqrt{\frac{N}{2}}(1 + o(1)) + O(1) \sim N\alpha_0\theta_m, \tag{13}$$

$$\mathbf{D}\nu_{k_m} \sim \frac{1}{2}N\alpha_0^2\theta_m^2. \tag{14}$$

Substituting (12-(14) into (11), we obtain

$$\mathbf{P}\left\{\frac{\mu_0(n_1 + \ldots + n_m) - Ne^{-\alpha_0\theta_m}}{\alpha_0\sqrt{N/2}} \leqslant x_m\theta_m, \ m = 1, \ldots, t\right\} \sim \mathbf{P}\left\{\frac{\nu_{k_m} - \mathbf{M}\nu_{k_m}}{\alpha_0\sqrt{N/2}} \leqslant x_m\theta_m, \ m = 1, \ldots, t\right\}.$$

It is seen from (13) that, under the conditions of Theorem 2,

$$\frac{k_1}{\sqrt{N}} \sim \alpha_0 \theta_1 \sqrt{N} \to \infty, \qquad \frac{k_m - k_{m-1}}{\sqrt{N}} \sim \alpha_0 (\theta_m - \theta_{m-1}) \sqrt{N} \to \infty,$$

as $N \longrightarrow \infty$. Therefore, the variables ν_{k_1} and $\nu_{k_m} - \nu_{k_{m-1}}$, for $m = 2, \ldots, t$, are asymptotically normal. Applying Lemma 1 to the random variables

$$\eta_1 = \frac{\nu_{k_1} - \mathbf{M}\nu_{k_1}}{\alpha_0 \sqrt{N/2}}, \qquad \eta_m = \frac{\nu_{k_m} - \nu_{k_{m-1}} - \mathbf{M}(\nu_{k_m} - \nu_{k_{m-1}})}{\alpha_0 \sqrt{N/2}}$$

$$\left(\mathbf{D}\eta_1 \sim \theta_1^2, \ \mathbf{D}\eta_m \sim \theta_m^2 - \theta_{m-1}^2, \ m = 2, \ldots, t\right),$$

we complete the proof of Theorem 2.

In a similar manner as in the right-hand intermediate domain, the random variables in (9) depend on the way $\alpha_0 \longrightarrow 0$. If $\alpha_0 \longrightarrow 0$ and $\alpha_0' \to 0$ in such a way that $|\ln \alpha_0 - \ln \alpha_0'| \to \infty$, $N\alpha_0^2 \to \infty$, and $N(\alpha_0')^2 \to \infty$, the finite-dimensional distributions of $\xi_{\alpha_0}(\theta)$ and $\xi_{\alpha_0'}(\theta)$ converge to finite-dimensional distributions of independent Gaussian processes.

§7. Further Results. References

The discussion in Sections 1-5 is based on Sevast'yanov [44]. The results given in Section 6 were obtained by Bolotnikov [6]. The convergence of $\mu_r(n) = \mu_r(n, N)$ to Poisson and Gaussian processes in the central, left-hand, and right-hand r-domains is proved in Bolotnikov [5]. We shall formulate here the main results given in [5], omitting the proofs.

Let us introduce the generating functions

$$F_{r,n_1,\ldots,n_t}(x_1, \ldots, x_t) =$$
$$= \sum_{k_1,\ldots,k_t} \mathbf{P}\{\mu_r(n_1) = k_1, \mu_r(n_1 + n_2) = k_2, \ldots$$
$$\ldots, \mu_r(n_1 + n_2 + \ldots + n_t) = k_t\} x_1^{k_1} x_2^{k_2} \ldots x_t^{k_t} \quad (1)$$

and

$$\Phi_{N,r}(z;\ x) = \Phi_{N,r}(z_1, \ldots, z_t;\ x_1, \ldots, x_t) =$$
$$= \sum_{n_1,\ldots,n_t=0}^{\infty} F_{r,n_1,\ldots,n_t}(x_1, \ldots, x_t) \prod_{m=1}^{t} \frac{(Nz_m)^{n_m}}{n_m!}. \qquad (2)$$

Theorem 1. *The generating function* $\Phi_{N,r}(z;\ x)$ *is equal to* $[\Phi_r(z;\ x)]^N$, *where*

$$\Phi_r(z,\ x) = b_1 - \sum_{k=1}^{t} a_k(b_{k+1} - b_{k+2}) +$$
$$+ \sum_{k=1}^{t} \sum_{s=k+1}^{t+1} x_k x_{k+1} \ldots x_{s-1}(a_k - a_{k-1})(b_s - b_{s-1})$$

and

$$a_k = \frac{(z_1 + z_2 + \ldots + z_k)^r}{r!}, \qquad k = 1, 2, \ldots, t, \qquad a_0 = 0,$$
$$b_m = e^{z_m + z_{m+1} + \ldots + z_t}, \quad m = 1, 2, \ldots, t;\ b_{t+1} = 1,\ b_{t+2} = 0.$$

In the right-hand r-domain, we introduce the time θ by

$$\frac{n}{N} - \ln N - r \ln \ln N + \ln r! = -\ln \theta. \qquad (3)$$

This time is "reciprocal," i.e., the larger n, the smaller θ. Let

$$\frac{n_1 + n_2 + \ldots + n_m}{N} - \ln N - r \ln \ln N + \ln r! =$$
$$= -\ln \theta_m + o(1), \qquad (4)$$
$$m = 1,\ 2,\ \ldots,\ t.$$

Theorem 2. *If* n_m, $N \longrightarrow \infty$ *in such a way that* $\theta_1, \ldots, \theta_m$ *in* (4) *are constants, the random variables*

$$\mu_r(n_1+\ldots+n_t),\ \mu_r(n_1+\ldots+n_m)-\mu_r(n_1+\ldots+n_{m+1}),$$
$$m=1, 2, \ldots, t-1,$$

are asymptotically independent and their distributions converge respectively to Poisson distributions with parameters θ_t *and* $\theta_m-\theta_{m+1}$ *for* $m=1, \ldots, t-1$.

In the central domain, we introduce the time $\theta=n/N$. From Theorem 1.1.1 and the generating function (1) we have, as $n, N\longrightarrow\infty$,

$$\mathbf{M}\mu_r(n)\sim N\frac{\theta^r}{r!}e^{-\theta}, \quad \theta=\frac{n}{N},$$

$$\mathbf{Cov}\,(\mu_r(n), \mu_r(n'))\sim$$
$$\sim N\left[\frac{\theta^r}{r!}e^{-\theta'}-\frac{\theta^r\theta'^r}{(r!)^2}e^{-\theta-\theta'}\left(1+\frac{(r-\theta)(r-\theta')}{\theta'}\right)\right],$$
$$\frac{n'}{N}=\theta'\geqslant\frac{n}{N}=\theta.$$

We introduce the process

$$\xi_r(\theta)=\frac{\mu_r(n)-\mathbf{M}\mu_r(n)}{\sqrt{N}}, \quad \frac{n}{N}\leqslant\theta<\frac{n+1}{N}.$$

Theorem 3. *If* $n/N=\theta+o(1)$ *as* $n, N\rightarrow\infty$, *the finite-dimensional distributions of the process* $\xi_r(\theta)$ *converge to finite-dimensional distributions of the Gaussian process* $\eta_r(\theta)$ *with correlation function*

$$\frac{\theta^r}{r!}e^{-\theta'}-\frac{\theta^r\theta'^r}{(r!)^2}e^{-\theta-\theta'}\left[1+\frac{(r-\theta)(r-\theta')}{\theta'}\right], \quad \theta'\geqslant\theta.$$

In the left-hand r-domain, $n, N\longrightarrow\infty$ in such a way that n^r/N^{r-1} is bounded. Let us introduce a new time $\theta=\dfrac{n^r}{r!N^{r-1}}$.

Theorem 4. *If* $r\geqslant 2$ *is fixed and* n_k, $N\longrightarrow\infty$ *in such a way that*

$$\frac{(n_1+\ldots+n_k)^r}{r!N^{r-1}}=\theta_k+o(1),$$

$$k = 1, 2, \ldots, r, \quad \theta_0 = 0,$$

then the random variables

$$\mu_r(n_1),\ \mu_r(n_1+\ldots+n_{k+1})-\mu_r(n_1+\ldots+n_k),$$
$$k=1, \ldots, t-1, \qquad (5)$$

are asymptotically independent and their distributions converge to Poisson distributions with parameters θ_1 *and* $\theta_{k+1}-\theta_k$, *for* $k=1, \ldots, t-1$, *respectively.*

If $r=1$, in the left-hand 1-domain we have the following limit theorem:

Theorem 5. *If* n_k, $N \longrightarrow \infty$ *in such a way that*

$$\frac{(n_1 + \ldots + n_k)^2}{2N} = \theta_k + o(1), \quad k = 1, \ldots, t, \quad \theta_0 = 0,$$

the random variables

$$\frac{n_1 - \mu_1(n_1)}{2}, \quad \frac{n_k + \mu_1(n_1 + \ldots + n_{k-1}) - \mu_1(n_1 + \ldots + n_k)}{2}$$
$$k = 2, \ldots, t$$

are asymptotically independent and their distributions converge to Poisson distributions with parameters θ_1 *and* $\theta_k-\theta_{k-1}$, *for* $k=2, \ldots, t$, *respectively.*

In Bolotnikov [7], the results given in Sections 3-5 are extended to the random variable $\mu_0(n)$ in a polynomial scheme of allocation.

In Hafner [76], [77], and [78], it is shown that several different configurations formed by particles occupying the nodes of an infinite lattice in a finite-dimensional space converge to a Poisson or Gaussian process.

Chapter V

THE EMPTY CELL TEST AND ITS GENERALIZATIONS

§1. The Empty Cell Test

The empty cell test was introduced in Section 1, Chapter 1 for testing a statistical hypothesis H_0 that a sample consisting of independent observations $x_1, x_2, \ldots, x_n$ is taken from the continuous distribution $F(x)$. This test is based on the statistic μ_0 defined as follows. Let us choose the points $z_0 = -\infty < z_1 < z_2 < \ldots < z_{N-1} < z_N = \infty$ so that

$$a_k = F(z_k) - F(z_{k-1}) = \frac{1}{N}, \quad k = 1,2, \ldots, N. \tag{1}$$

The statistic μ_0 is equal to the number of intervals $(z_{k-1}, z_k]$, into which no observation fell. The test is as follows: if $\mu_0 \leqslant C$, the hypothesis H_0 is adopted; if $\mu_0 > C$, it is rejected. The error of the first kind, i.e., the probability of rejecting H_0 when this hypothesis is true, is equal to

$$\gamma(N) = \mathbf{P}\{\mu_0(n, N) > C\}, \tag{2}$$

where $\mu_0(n, N)$ is the number of empty boxes in a polynomial scheme with the probabilities a_k defined by (1). For large values of n, N, it is natural to apply limit theorems in calculating $\gamma(N)$. By Theorem 1.3.2 and the formulas given in (1.1.14) and (1.1.15), in the central domain the variable $\mu_0(n, N)$ is asymptotically normal with parameters $(NA_0, \sigma_0\sqrt{N})$, where

$$A_0 = e^{-\alpha}, \quad \sigma_0^2 = e^{-\alpha}[1-(1+\alpha)e^{-\alpha}], \quad \alpha = \frac{n}{N}. \tag{3}$$

We define the variables u_t, for $0<t<1$, by

$$t = \frac{1}{\sqrt{2\pi}} \int_{u_t}^{\infty} e^{-x^2/2}dx.$$

Let us set

$$C = NA_0 + u_\gamma \sigma_0 \sqrt{N}, \quad 0<\gamma<1. \tag{4}$$

If the hypothesis H_0 is true, then by Theorem 1.3.2 for the variable $\gamma(N)$ defined by (2) with a constant C of the form (4), we have $\gamma(N) \longrightarrow \gamma$ as $N \longrightarrow \infty$.

If a simple competing hypothesis H_1 is true (the sample is taken from a continuous distribution $G(x) \not\equiv F(x)$), the probabilities $a_k = G(z_k) - G(z_{k-1})$, for $k=1, 2, \ldots, N$, of getting into the intervals (z_{k-1}, z_k) are not necessarily equal to $1/N$. By Theorem 3.4.2, the variable μ_0 is asymptotically normal with parameters $(NA_1, \sigma_1\sqrt{N})$, where A_1 and σ_1^2 are defined by (3.1.17) and (3.1.18). Thus, we have arrived at the problem of testing of two simple hypotheses

$$\begin{aligned} &H_0 : a_k = \frac{1}{N}, \quad k = 1, 2, \ldots, N, \\ &H_1 : a_k = a_k^* \quad \left(a_k^* \neq \frac{1}{N}\right), \quad k = 1, 2, \ldots, N, \end{aligned} \tag{5}$$

regarding the probabilities of outcomes in a polynomial scheme depending on the number of emply cells μ_0.

If the function $F(x)$ is continuous, then transforming the observations x_k into $y_k=F(x_k)$, we obtain the sample from a uniform distribution. Therefore, we can assume that $F(x)\equiv x$, $0\leqslant x\leqslant 1$. If the density $g(x)$ of the distribution $G(x)$ is several times differentiable, with H_1 we have for the probabilities a_k^*

$$a_k^* = \frac{1}{N}\, g\left(\frac{k}{N}\right) - \frac{1}{2N^2}\, g'\left(\frac{k}{N}\right) + O\left(\frac{1}{N^3}\right), \quad k = 2, \ldots, N-1,$$

$$a_1^* = G\left(\frac{1}{N}\right), \quad a_N = 1 - G\left(1 - \frac{1}{N}\right). \tag{6}$$

In the case where the probabilities are of the form (6), it follows immediately from (3.1.17) and (3.1.18) that

$$\lim_{N\to\infty} A_1 = \int_0^1 e^{-\alpha g(x)}dx,$$

$$\lim_{N\to\infty} \sigma_1^2 = \int_0^1 e^{-\alpha g(x)}dx - \int_0^1 e^{-2\alpha g(x)}dx -$$

$$- \alpha\left(\int_0^1 g(x)\, e^{-\alpha g(x)}dx\right)^2, \tag{7}$$

if $n/N\to\alpha=\text{const}$. Denote by $\delta(N)$ the error of the second kind (the probability of accepting the hypothesis H_0 when the hypothesis H_1 is true). If the hypothesis H_1 is true, then

$$\delta(N) = \mathbf{P}\{\mu_0 \leqslant C\} = \mathbf{P}\left\{\frac{\mu_0 - NA_1}{\sigma_1\sqrt{N}} < \frac{A_0 - A_1}{\sigma_1}\sqrt{N} + u_\gamma\frac{\sigma_0}{\sigma_1}\right\}. \tag{8}$$

If $A_0\neq A_1$ it can be easily verified that $A_0<A_1$. Then by Theorem 3.4.2, $\delta(N)\longrightarrow 0$ as $N\longrightarrow\infty$ since

$$\frac{\sqrt{N}(A_0 - A_1)}{\sigma_1} + u_\gamma\frac{\sigma_0}{\sigma_1} \to -\infty. \tag{9}$$

The case $\delta(N) \longrightarrow \delta > 0$ is of interest. From (8) and (9) it follows that this case is possible if $A_1 - A_0 \approx c/\sqrt{N}$ (where c is a positive constant).

Theorem 1. *If the* $\overset{*}{a}_k$ *in* (5) *are defined by* (3.1.23), *then the errors* $\gamma(N)$ *and* $\delta(N)$ *of the first and second kind of the empty cell test with the constant* (4) *have limits* $\gamma > 0$ *and* $\delta > 0$ *respectively as* $N \longrightarrow \infty$, $\alpha = \frac{n}{N} \to \alpha_0$ *(where* $(0 < c_0 \leqslant \alpha \leqslant c_1 < \infty)$, *and*

$$u_\gamma + u_\delta = \frac{\alpha_0^2 b^2}{2\sigma_0} e^{-\alpha_0},$$

where σ_0, b^2 *are given by* (3) *and* (3.1.23).

PROOF. The convergence of $\gamma(N)$ to γ follows from the definition (4) of C and the asymptotic normality of $\mu_0(n, N)$ with parameters $(NA_0, \sigma_0\sqrt{N})$. Under the conditions of the theorem, by (3.1.24),

$$A_1 = A_0\left(1 + \frac{b_N^2}{2\sqrt{N}}\alpha^2\right) + O\left(\frac{1}{N^{3/4}}\right), \quad \sigma_1 = \sigma_0 + O\left(\frac{1}{\sqrt{N}}\right).$$

Therefore, instead of (9) we have

$$\lim_{N\to\infty}\left[\frac{\sqrt{N}(A_0 - A_1)}{\sigma_1} + u_\gamma \frac{\sigma_0}{\sigma_1}\right] = -\frac{\alpha_0^2 b^2}{2\sigma_0} e^{-\alpha_0} + u_\gamma. \qquad (10)$$

By virtue of (8), the right-hand side of (10) is u_δ. The theorem is proved.

Let us take δ as a function of α for fixed γ and b. Then

$$u_{\delta(\alpha)} = \Psi(\alpha)\frac{b^2}{2} - u_\gamma, \qquad (11)$$

where

$$\Psi(\alpha) = \frac{\alpha^2}{\sqrt{e^\alpha - 1 - \alpha}}. \qquad (12)$$

As α increases from $\alpha=0$ to $\alpha=1$, the function $\Psi(\alpha)$ increases. Therefore, $u_{\delta(\alpha)}$ increases and so does the power of the test $1-\delta(\alpha)$. The increase in $\alpha=n/N$ corresponds to the increase in the number of observations. However, for a large number of observations ($\alpha>1$), there are few empty cells under either hypothesis, a fact that decreases the power.

§2. Linear Statistic

In the previous section, we examined a test which distinguishes the hypotheses (1.5) and which was based on the statistic μ_0. This test can be naturally generalized if along with the statistic μ_0 a set of statistics μ_r is used for different r.

First we consider the linear test based on the statistic

$$\xi_r=c_{0r}\mu_0+c_{1r}\mu_1+\ldots+c_{rr}\mu_r. \tag{1}$$

We distinguish the hypotheses H_0 and H_1 (see (1.5)), when the $\overset{*}{a}_k$ are given by (3.1.23). Let us denote by $\mathbf{M}_i$ and $\mathbf{D}_i$ the mathematical expectation and the variance respectively, calculated from the distribution of the hypothesis H_i for $i=0, 1$.

Making use of formulas (3.1.24), we obtain easily the result that, as $N\longrightarrow\infty$,

$$\mathbf{M}_1\xi_r=N\sum_{k=0}^{r}c_{kr}p_k\left(1+\frac{b_N^2}{2\sqrt{N}}[(\alpha-k)^2-k]\right)+O\left(\sqrt[4]{N}\right),$$

$$\mathbf{D}_1\xi_r=N\sum_{k,l=1}^{r}c_{kr}c_{lr}\sigma_{kl}+O\left(\sqrt{N}\right), \tag{2}$$

where $p_k=\frac{\alpha^k}{k!}e^{-\alpha}$, σ_{kl}, and b_N^2 are given by (2.1.9) and (3.1.23). Setting $b_N^2=0$ in (2), we have

$$\begin{aligned} \mathbf{M}_0\xi_r &= N \sum_{k=0}^{r} c_{kr} p_k + O\left(\sqrt[4]{N}\right), \\ \mathbf{D}_0\xi_r &= N \sum_{k,\,l=0}^{r} c_{kr} c_{lr} \sigma_{kl} + O\left(\sqrt{N}\right). \end{aligned} \tag{3}$$

as $N \longrightarrow \infty$.

Let

$$C = N \sum_{k=0}^{r} c_{kr} p_r + u_\gamma \sqrt{N \sum_{k,\,l=0}^{r} c_{kr} c_{kl} \sigma_{kl}}. \tag{4}$$

We accept the hypothesis H_0 if

$$\xi_r \leqslant C, \tag{5}$$

and the hypothesis H_1 if

$$\xi_r > C. \tag{6}$$

Let us denote by $\gamma(N)$ and $\delta(N)$ the errors of the first and the second kinds, respectively, of this test. Using Theorems 2.2.3 and 3.5.2 and formulas (2) and (3) in the same way as in proving Theorem 1.1, we see that

$$\gamma(N) \longrightarrow \gamma, \quad \delta(N) \longrightarrow \delta$$

as $N \longrightarrow \infty$ and that

$$u_\gamma + u_\delta = \frac{b^2 \sum_{k=0}^{r} c_{kr} p_k \left[(\alpha - k)^2 - k\right]}{2\sqrt{\sum_{k,\,l=0}^{r} c_{kr} c_{lr} \sigma_{kl}}}. \tag{7}$$

We note that in calculating the errors of the first and the second kinds, it is more convenient to represent the expression in the denominator in the following form:

$$\sum_{k,l=0}^{r} c_{kr}c_{lr}\sigma_{kl} = \sum_{k=0}^{r} c_{kr}^2 p_k - \left(\sum_{k=0}^{r} c_{kr}p_k\right)^2 - \\ - \alpha\left[\left(\sum_{k=0}^{r} c_{kr}p_k\right) - \left(\sum_{k=0}^{r-1} c_{k+1,r}p_k\right)\right]^2.$$

Theorem 1. *Among the tests defined by* (5) *and* (6) *with the statistic* (1), *the asymptotically most powerful test, as* n, $N \to \infty$ *and* $\frac{n}{N} \to \alpha_0$, *where* $0 < \alpha_0 < \infty$, *is yielded by the statistic*

$$\xi_r = \sum_{k=0}^{r} \overset{*}{c}_{kr}\mu_k, \tag{8}$$

where

$$\overset{*}{c}_{kr} = 1 - \frac{2k}{\alpha_0} + \frac{k(k-1)}{\alpha_0^2} + \frac{\alpha_0 - r}{\alpha_0}\theta_r + \\ + \frac{\theta_r[\alpha_0(1+\theta_r) - k][\theta_r(\alpha_0 - r) + ((\alpha_0 - r)^2 + r)/\alpha_0]}{\alpha_0[1 - \theta_r(\alpha_0 - r - 1 + \alpha_0\theta_r)]} \tag{9}$$

$$\theta_r = \frac{\alpha_0^r}{r!}\left(\sum_{k=r+1}^{\infty}\frac{\alpha_0^k}{k!}\right)^{-1}.$$

PROOF. We need to show that the power $1-\delta$ given by (7) is the maximal one when $c_{kr} = \overset{*}{c}_{kr}$. The largest value of the power corresponds to the largest value of u_δ. Since multiplication of each c_{kr} by the same constant does not change the value of (7), we can replace our problem with the following problem: On the $(r+1)$-dimensional ellipsoid

$$\sum_{k,l=0}^{r} \sigma_{kl}x_kx_l = 1, \tag{10}$$

find the point $(x_0, x_1, \ldots, x_r)$ at which the function

$$y(x_0, \ldots, x_r) = \sum_{k=1}^{r} x_k A_k, \qquad A_k = p_k[(\alpha_0 - k)^2 - k],$$

attains its maximum. Using the method of the Lagrange multipliers to find the conditional extremum, we obtain the system of equations

$$\sum_{l=0}^{r} \sigma_{kl} x_l = -\frac{A_k}{2\lambda}, \qquad k = 0, 1, \ldots, r,$$

for the extremal points $(x_0, \ldots, x_r)$. From this we get

$$x_l = -\frac{1}{2\lambda B^2} \sum_{k=0}^{r} B_{kl} A_k, \qquad l = 0, 1, \ldots, r, \tag{11}$$

where B^2 is the determinant of the matrix $\|\sigma_{kl}\|$ and B_{kl} is the cofactor of the element σ_{kl}. Substituting (11) into (10), we obtain for λ a quadratic equation with two roots λ_1 and λ_2 such that $\lambda_1 = -\lambda_2$. Therefore, we have two extremal points whose coordinates are of opposite signs. The maximum of $y(x_0, \ldots, x_r)$ is attained at one of these points, the minimum at the other. Using the explicit formulas for B^2 and B_{kl} (namely, 2.2.5), (2.2.8), (2.2.9)), we can prove the collinearity of the vectors $(x_0, \ldots, x_r)$ and $(\overset{*}{c}_{0r}, \ldots, \overset{*}{c}_{rr})$ whose components are given by (11) and (9). To complete the proof of the theorem, we only need to show that the value of the function $y(x_0, \ldots, x_r)$ is maximum when $x_k = \overset{*}{c}_{kr}$ for $k = 0, \ldots, r$. To this end, it suffices to verify that $y(\overset{*}{c}_{0r}, \ldots, \overset{*}{c}_{rr}) > 0$, since in our problem the minimum and the maximum are of opposite signs. We point out that, if $x_k = \overset{*}{c}_{kr}$ for $k = 0, 1, \ldots, r$, the function $y(x_0, \ldots, x_r)$ does not vanish for any value of α_0. To see this, note that, if there is such an α_0, the maximum and the minimum of the function $y(x_0, \ldots, x_r)$ must be zero for this α_0, which is not possible since our linear function is not identically equal to zero. For sufficiently large α_0, we have $y(\overset{*}{c}_{0r}, \ldots, \overset{*}{c}_{rr}) > 0$; from this, according to the remark made above, it follows that $y(\overset{*}{c}_{0r}, \ldots, \overset{*}{c}_{rr})$ is positive for any α_0 since this function is continuous with respect to α_0. Therefore, the maximum is attained at the point $x_k = \overset{*}{c}_{kr}$ for $k=0, 1, \ldots, r$. The theorem is proved.

Analyzing the formulas given in (9) for small α_0 as well as large α_0, we see that the best linear test is defined by the formula

$$\xi_r = C^2_{r+2}\mu_0 + C^2_{r+1}\mu_1 + \ldots + 3\mu_{r-1} + \mu_r$$

for small α_0 and by

$$\xi_r = \mu_0 + \mu_1 + \ldots + \mu_r$$

for large α_0.

§3. The Optimal Statistic

We shall show that the test found in the previous section is optimal among all tests based on variables $\mu_0, \mu_1, \ldots, \mu_r$.

Theorem 1. *Among all tests distinguishing the hypotheses* (1.5), *where the* $\overset{*}{a}_k$ *are defined by* (3.1.23), *on the basis of* $\mu_0, \mu_1, \ldots, \mu_r$, *the linear statistic* (2.8) *is the asymptotically most powerful as* $n, N \to \infty$ *and* $\alpha = \frac{n}{N} \to \alpha_0$, *where* $0 < \alpha_0 < \infty$.

Proof. Under the hypothesis H_i (for $i=0, 1$), according to Theorems 2.1.1 and 3.5.3, we have for the probabilities

$$P_N^{(i)}(k_0, k_1, \ldots, k_r) = \mathbf{P}\{\mu_0 = k_0, \ldots, \mu_r = k_r\}$$

the asymptotic formula

$$P_N^{(i)}(k_0, \ldots, k_r) = \frac{1+o(1)}{\sqrt{(2\pi N)^{r+1}}} \exp\left(-\frac{1}{2}\sum_{k,l=0}^{r} B_{kl} u_k^{(i)} u_l^{(i)}\right), \quad (1)$$

as $n, N \longrightarrow \infty$, where

$$u_l^{(i)} = \frac{k_l - NA_{il}}{B\sqrt{N}}, \quad |u_l^i| \leqslant C < \infty,$$
$$A_{1l} = p_l\left(1 + \frac{b_N^2}{2\sqrt{N}}[(\alpha - l)^2 - l]\right), \quad A_{0l} = p_l, \tag{2}$$

$p_l = \frac{\alpha^l}{l!}e^{-\alpha}$ and B, B_{kl}, and b_N^2 are given by (2.2.5), (2.2.8), (2.2.9), and (3.1.23). We note that from (2) we get

$$\lim_{N\to\infty} \sqrt{N}(A_{1l} - A_{0l}) = \frac{b^2}{2}[(\alpha_0 - l)^2 - l]. \tag{3}$$

By the Neyman-Pearson lemma, the random variable

$$\eta_n = \ln \frac{P_N^{(1)}(\mu_0, \mu_1, \ldots, \mu_r)}{P_N^{(0)}(\mu_0, \mu_1, \ldots, \mu_r)} \tag{4}$$

is the statistic corresponding to the most powerful test distinguishing the two simple hypotheses H_0 and H_1.

Using (1) and the formulas

$$v_k^{(0)} = \frac{\mu_k - NA_{0k}}{B\sqrt{N}}, \quad v_k^{(1)} = \frac{\mu_k - NA_{1k}}{B\sqrt{N}} = v_k^{(0)} + \frac{\sqrt{N}}{B}(A_{0k} - A_{1k}),$$

we obtain from (4) the result that

$$\eta_N = \frac{1}{2}\sum_{k,l=0}^{r} B_{kl}\Big[v_k^{(0)}\sqrt{N}(A_{1l} - A_{0l}) +$$
$$+ v_l^{(0)}\sqrt{N}(A_{1k} - A_{0k}) - \frac{N}{B^2}(A_{1k} - A_{0k})(A_{1l} - A_{0l})\Big] + \nu_N. \tag{5}$$

If $|v_k^{(i)}| \leqslant C < \infty$ for $i = 0, 1$, then ν_N approaches zero. Let us show that ν_N converges in probability to zero under the hypothesis H_i (for i=0, 1). We partition the set R of values of the vector $(\mu_0, \mu_1, \ldots, \mu_r)$ into two parts: R_ε and $R_{1-\varepsilon} = R \setminus R_\varepsilon$;

we choose $R_{1-\varepsilon}$ so that, for $(\mu_0, \ldots, \mu_r) \in R_{1-\varepsilon}$, the inequality $|v_k^{(i)}| < C_i$ is satisfied, C_i being subject to the condition

$$\mathbf{P}\{(\mu_0, \ldots, \mu_r) \in R_{1-\varepsilon}\} \geqslant 1-\varepsilon. \tag{6}$$

This is possible by virtue of (1). Since (6) is satisfied and the value $\nu_N \longrightarrow 0$ in $R_{1-\varepsilon}$ as $N \longrightarrow \infty$, it follows that for any $\varepsilon > 0$ and $\delta > 0$ there is N_0 such that, for all $N > N_0$,

$$\mathbf{P}\{|\nu_N| < \delta\} > 1-\varepsilon.$$

This implies that ν_N converges in probability to zero. Noting (3), we put (5) in the form

$$\eta_N = \sum_{l=0}^{r} c_l v_l^{(0)} + D + \tilde{\nu}_N,$$

where

$$c_l = \sum_{k=0}^{r} B_{kl} \frac{b^2}{2} [(\alpha_0 - k)^2 - k],$$
$$D = -\frac{b^4}{4B^2} \sum_{k,l=0}^{r} [(\alpha_0 - l)^2 - l]\,[(\alpha_0 - k)^2 - k], \tag{7}$$

and $\tilde{\nu}_N$ converges in probability to zero under either hypothesis H_i $(i=0, 1)$. From this it follows that the limit distributions of the variables η_N and $\eta_N - \tilde{\nu}_N$ coincide under either hypothesis H_i (for $i=0, 1$); i.e., the linear statistic $\eta_N - \tilde{\nu}_N$ has a distribution coinciding asymptotically with the distribution of the statistic of the most powerful test η_N. Since $\eta_N - \tilde{\nu}_N$ depends linearly on μ_0, $\mu_1, \ldots, \mu_r$, the power of the test based on the above statistic must coincide with the power of the optimal linear test. The theorem is proved.

Comparing (7) and (2.11), we see that c_l and x_l are proportional. Thus, we see immediately that the statistic $\eta_N - \tilde{\nu}_N$ is a linear function of the statistic of the optimal linear test.

§4. Further Results. References

The empty cell test is widely known (see, for example, David [70], Okamoto [89], Kitabatake [84], Sevast'yanov [45], Chistyakov [55]) and explained in standard textbooks ([53], Chapter 14). For calculating errors of the first kind with this test, the theorems on asymptotic normality of μ_0 are used in an equiprobable scheme under various assumptions (see Okamoto [89], Weiss [98], Rényi [91]; Section 1.6). The power of the test is computed via Theorem 3.4.2, due to Chistyakov [55]. These results can be used for large values of n and N. For values of n and N in the intervals

$$2 \leqslant N \leqslant 10, \qquad 5 \leqslant n \leqslant 50,$$

Csorgo and Guttman [68] have drawn up a table of values C for errors of the first kind equal to 0.01 and 0.05. Csorgo and Guttman [67] and [68] and Wilks [99] study the two-sample test of empty boxes. This test is as follows. The observations of the first sample form a partition of the real line into cells. Next, the number of cells containing no observations of the second sample is found. From the number of empty cells, one decides whether the two samples are distributed according to the same law or not.

Theorems 2.1 and 3.1 are obtained by Viktorova and V. P. Chistyakov [13]. The investigation of statistical tests is naturally related to that of linear functions of μ_r. Ivchenko [19] studied the sums $\xi_m = \mu_0 + \mu_1 + \ldots + \mu_m$ in an equiprobable scheme. Harris and Park [79] studied the variable $W = \sum_{r=0}^{n} \omega_r \mu_r(n, N)$ to obtain the following results.

Theorem 1. *Suppose that* $\frac{n}{N} \to \alpha > 0$. *If for some* $\beta > \alpha$

$$\sum_{i > N^{1/5}} \frac{\omega_i^2 \beta^i}{i!} = o\left(\frac{1}{N}\right),$$

then, as $n,\ N \longrightarrow \infty$,

$$m = \mathbf{M}W \sim Ne^{-\alpha} \sum_{i=0}^{\infty} \frac{w_i \alpha^i}{i!},$$

$$\sigma^2 = \mathbf{D}W \sim$$

$$\sim Ne^{-2\alpha} \sum_{i=0}^{\infty} \sum_{j=0}^{\infty} \frac{\alpha^{i+j}}{i!j!} \left[w_i w_j \left(i + j - \frac{ij}{\alpha} - \alpha - 1 \right) + w_i^2 \right].$$

Theorem 2. *If the conditions of Theorem 1 are satisfied and if* $\frac{\sigma^2}{N} \rightarrow d^2 > 0$, *then* W *is asymptotically normal with parameters* (m, σ).

Choosing coefficients w_r in a particular way, one can obtain statistics of the empty cell test of μ_0, the maximal likelihood ratio

$$\lambda = \sum_{k=0}^{N} n_k \ln n_k - \ln \frac{n}{N} = \sum_{r=0}^{n} (r \ln r) \mu_r - n \ln \frac{n}{N}$$

and

$$\chi^2 = \sum_{k=1}^{N} \frac{\left(n_k - \frac{n}{N} \right)^2}{n/N} = \frac{n}{N} \sum_{r=0}^{n} r^2 \mu_r - n.$$

Here n_k is the number of shots in the kth cell. The limit distributions of the statistic χ^2 in a polynomial scheme, as $N \rightarrow \infty$, are investigated in Tumanyan [51], [52], and Medvedev [36]. The local normal theorem for the sums of the squares of the frequencies, as $N \longrightarrow \infty$, is due to Kolchin [32].

Chapter VI

ALLOCATIONS WITH THE NUMBER OF PARTICLES RANDOMIZED

§1. Moments and a General Limit Theorem

Consider an equiprobable scheme of allocation which is more complicated than hitherto by replacing the number n of particles by a random variable ν. Let us assume that first we are given a probability distribution $\mathbf{P}\{\nu=n\}$. We examine the equiprobable scheme of allocation of n particles into N cells, conditioned on $\nu=n$. Let μ_r denote the number of cells containing exactly r particles.

Let $\varphi(\cdot)$ denote the generating function for the number of particles ν:

$$\varphi(x) = \sum_{n=0}^{\infty} \mathbf{P}\{\nu = n\} x^n. \tag{1}$$

The factorial moments $\mathbf{M}\mu_r^{[R]}$ are readily expressed via the derivatives of $\varphi(x)$.

Theorem 1. *If* ν *has the generating function* (1), *for any* *integers* $r \geqslant 0$ *and* $k \geqslant 0$,

$$\mathbf{M}\mu_r^{[k]} = \frac{N^{[k]}}{N^{rk}(r!)^k}\varphi^{(rk)}\left(1-\frac{k}{N}\right). \tag{2}$$

PROOF. We established in Theorem 2.1.3 that in an equiprobable scheme of allocation, for any integers $r \geqslant 0$ and $k \geqslant 0$,

$$\mathbf{M}\mu_r^{[k]}(n) = \frac{N^{[k]}n^{[rk]}}{N^{rk}(r!)^k}\left(1-\frac{k}{N}\right)^{n-rk}. \tag{3}$$

Since in our scheme

$$\mathbf{M}\{\mu_r^{[k]} \mid \nu = n\} = \mathbf{M}\mu_r^{[k]}(n), \tag{4}$$

we obtain from (3) and (4)

$$\mathbf{M}\mu_r^{[k]} = \mathbf{M}\left[\mathbf{M}\mu_r^{[k]} \mid \nu = n\right] =$$

$$= \sum_{n=0}^{\infty}\mathbf{P}\{\nu = n\}\frac{N^{[k]}n^{[rk]}}{N^{rk}(r!)^k}\left(1-\frac{k}{N}\right)^{n-rk} = \frac{N^{[k]}}{N^{rk}(r!)^k}\varphi^{(rk)}\left(1-\frac{k}{N}\right).$$

Further, we assume that the probability distribution of the random variable ν depends in some way or other on N. Letting $N \to \infty$, we shall prove some limit theorems. In this section, we prove a general limit theorem relevant to the case in which the distribution ν/N has a limit.

Theorem 2. *If the distribution function of the random variable* ν/N *converges weakly to the distribution function* $\Psi(x)$ *as* $N \to \infty$, *then the probability distribution of the random variable* μ_r/N *converges weakly, as* $N \to \infty$, *to the probability distribution of the random variable* $p_r(\xi) = \frac{\xi^r}{r!}e^{-\xi}$, *where* $\mathbf{P}\{\xi \leqslant x\} = \Psi(x)$.

PROOF. Let $\psi_N(s) = \varphi(e^{-s/N})$ denote the Laplace transform of (the distribution of) the random variable ν/N. By the hypothesis of the theorem,

$$\psi_N(s) \longrightarrow \psi(s), \tag{5}$$

where $\psi(s) = \int_0^\infty e^{-sx}\, d\Psi(x)$. To prove this theorem, we invoke the method of moments. It follows from (5) that for $s>0$

$$(-1)^l \mathbf{M}\left(\frac{\nu}{N}\right)^l e^{-s\nu/N} = \psi_N^{(l)}(s) \to \psi^{(l)}(s) \tag{6}$$

as $N \longrightarrow \infty$ for any $l>0$. To see this, we obtain from (5) the result that, for any bounded continuous function $f(x)$,

$$\int f(x)\, d\Psi_N(x) \to \int f(x)\, d\Psi(x).$$

Setting $f(x) = x^l e^{-sx}$, we arrive at (6).

Let us prove that, for $k=0, 1, 2, \ldots,$

$$\lim_{N\to\infty} \mathbf{M}\left(\frac{\mu_r}{N}\right)^k = \frac{(-1)^{kr}\, \psi^{(kr)}(k)}{(r!)^k}. \tag{7}$$

Since, for nonnegative integers x,

$$x^k - C_k^2 x^{k-1} \leqslant x^{[k]} \leqslant x^k, \tag{8}$$

we have:

$$\lim_{N\to\infty} \mathbf{M}\left(\frac{\mu_r}{N}\right)^k = \lim_{N\to\infty} \frac{\mathbf{M}\mu_r^{[k]}}{N^k}.$$

Hence (in view of (2) and (7)) we need only to prove that, for any k,

$$\lim_{N\to\infty} \frac{1}{N^{rk}(r!)^k}\, \varphi^{(rk)}\left(1 - \frac{k}{N}\right) = \frac{(-1)^{kr}\, \psi^{(kr)}(k)}{(r!)^k}. \tag{9}$$

Let $l=kr$. We have

$$\frac{1}{N^l}\varphi^{(l)}\left(1-\frac{k}{N}\right)=\mathbf{M}\frac{\nu^{[l]}}{N^l}\left(1-\frac{k}{N}\right)^{\nu-l}.$$

Since $1-x<e^{-x}$ for all $x\geqslant 0$ and $e^{-x^2-x}\leqslant 1-x$ for $0\leqslant x\leqslant 1/2$, we can write, using (8),

$$\left[\left(\frac{\nu}{N}\right)^l-\frac{C_l^2}{N}\left(\frac{\nu}{N}\right)^{l-1}\right]e^{-\left(k+\frac{k^2}{N}\right)\frac{\nu}{N}}\leqslant\frac{\nu^{[l]}}{N^l}\left(1-\frac{k}{N}\right)^{\nu-l}\leqslant$$
$$\leqslant\left(\frac{\nu}{N}\right)^l e^{-k\frac{\nu}{N}}\left(1-\frac{k}{N}\right)^{-l},$$

and hence,

$$\mathbf{M}\left(\frac{\nu}{N}\right)^l e^{-\left(k+\frac{k^2}{N}\right)\frac{\nu}{N}}-\frac{C_l^2}{N}\mathbf{M}\left(\frac{\nu}{N}\right)^{l-1}e^{-\left(k+\frac{k^2}{N}\right)\frac{\nu}{N}}\leqslant$$
$$\leqslant\frac{1}{N^l}\varphi^{(l)}\left(1-\frac{k}{N}\right)\leqslant\mathbf{M}\left(\frac{\nu}{N}\right)^l e^{-k\frac{\nu}{N}}\left(1-\frac{k}{N}\right)^{-l}. \quad (10)$$

Since $\frac{k^2}{N}\to 0$ and $\frac{kl}{N}\to 0$ as $N\to\infty$, if we let N approach ∞ in (10), we have for any $\delta>0$

$$\lim_{N\to\infty}\mathbf{M}\left(\frac{\nu}{N}\right)^l e^{-(k+\delta)\frac{\nu}{N}}\leqslant$$
$$\leqslant\lim_{N\to\infty}\frac{1}{N^l}\varphi^{(l)}\left(1-\frac{k}{N}\right)\leqslant\lim_{N\to\infty}\mathbf{M}\left(\frac{\nu}{N}\right)^l e^{-k\frac{\nu}{N}+\delta}.$$

We obtain (7) from (6) and the relation

$$\mathbf{M}\left(\frac{\nu}{N}\right)^l e^{-k\frac{\nu}{N}}=(-1)^l\psi_N^{(l)}(k).$$

We shall prove next that the limiting moments given by (7)

correspond to the limit distribution in Theorem 2. For this we have only to note that, using these moments, we can construct the Laplace transform

$$\sum_{k=0}^{\infty} \frac{(-1)^k s^k (-1)^{kr}}{k!\,(r!)^k} \psi^{(kr)}(k) =$$

$$= \sum_{k=0}^{\infty} \frac{(-1)^k s^k}{k!\,(r!)^k} \int_0^{\infty} x^{kr} e^{-kx} d\Psi(x) = \int_0^{\infty} \exp\left\{-s\frac{x^r}{r!}e^{-x}\right\} d\Psi(x).$$

The theorem we have just proved can be easily extended to the multivariate case.

Theorem 3. *If the distribution function ν/N converges weakly to $\Psi(x)$ as $N \to \infty$, the joint distribution of the random variables $\frac{\mu_{r_1}}{N}, \ldots, \frac{\mu_{r_t}}{N}$ converges weakly, as $N \to \infty$, to the joint distribution of the random variables $p_{r_1}(\xi), \ldots, p_{r_t}(\xi)$, where $p_r(x) = \frac{x^r}{r!} e^{-x}$ and $\mathbf{P}\{\xi \leqslant x\} = \Psi(x)$.*

PROOF. Starting from the formula for the multidimensional factorial moment (2.1.36)

$$\mathbf{M}\mu_{r_1}^{[k_1]}(n, N) \ldots \mu_{r_t}^{[k_t]}(n, N) =$$

$$= \frac{N^{[k]} n^{[(r,k)]}}{(r_1!)^{k_1} \ldots (r_t!)^{k_t} N^{(r,k)}} \left(1 - \frac{k}{N}\right)^{n-(r,k)},$$

where $k = k_1 + \ldots + k_t$ and $(r, k) = r_1 k_1 + \ldots + r_t k_t$, we can easily obtain a formula analogous to (2) for multidimensional moments:

$$\mathbf{M}\mu_{r_1}^{[k_1]} \ldots \mu_{r_t}^{[k_t]} = \frac{N^{[k]}}{N^{(r,k)} (r_1!)^{k_1} \ldots (r_t!)^{k_t}} \varphi^{((r,k))}\left(1 - \frac{k}{N}\right).$$

Just as in Theorem 2, we prove that, for arbitrary integers $k_i \geqslant 0$,

$$\lim_{N\to\infty} \mathbf{M}\left(\frac{\mu_{r_1}}{N}\right)^{k_1}\ldots\left(\frac{\mu_{r_t}}{N}\right)^{k_t} = \lim_{N\to\infty} \frac{\mathbf{M}\mu_{r_1}^{[k_1]}\ldots\mu_{r_t}^{[k_t]}}{N^k} =$$

$$= \lim_{N\to\infty} \frac{1}{N^{(r,k)}(r_1!)^{k_1}\ldots(r_t!)^{k_t}}\varphi^{(r,k)}\left(1-\frac{k}{N}\right) =$$

$$= \frac{(-1)^{(k,r)}}{(r_1!)^{k_1}\ldots(r_t!)^{k_t}}\psi^{((k,r))}(k).$$

The assertion of the theorem follows from the fact that

$$\sum_{k_1,\ldots,k_t} \frac{(-1)^k s_1^{k_1}\ldots s_t^{k_t}(-1)^{(k,r)}}{k_1!\ldots k_t!}\psi^{((k,r))}(k) =$$

$$= \int_0^\infty \exp\left\{\left(-s_1\frac{x^{r_1}}{r_1!}-\ldots-s_t\frac{x^{r_t}}{r_t!}\right)e^{-x}\right\}d\Psi(x).$$

The following corollary to Theorem 3 is of some interest. If

$$\lim_{N\to\infty} \mathbf{P}\left\{\frac{\nu}{N}\leqslant x\right\} = \Psi_r(x) = \frac{1}{r!}\int_0^x y^r e^{-y}dy, \tag{11}$$

then $(\mu_0+\mu_1+\ldots+\mu_r)/N$ has, in the limit as $N\to\infty$, a uniform distribution on the interval [0, 1]. In fact, by Theorem 3 $(\mu_0+\mu_1+\ldots+\mu_r)/N$ will have in the limit the same distribution as the random variable $\eta_r = \sum_{k=0}^{r} p_r(\xi)$ has, where $\mathbf{P}\{\xi\leqslant x\} = \Psi_r(x)$. Integrating the right-hand side of (11) by parts, we get $1-\Psi_r(x) = \sum_{k=0}^{r} p_k(x)$; hence $1-\eta_r = \Psi_r(\xi)$. From this it follows that the random variable $1-\eta_r$—hence η_r—is distributed uniformly on (0, 1).

§2. A Normal Limit Theorem

In this section, it will be assumed that, with the probability approaching one as $N\to\infty$, the random variable ν lies in the central domain. We restrict ourselves to the case where ν is asymptotically normal.

Theorem 1. *If the random variable ν is asymptotically normal as $N\to\infty$ with parameters $(aN, \sigma\sqrt{N})$, where $a>0$, then μ_r will also be asymptotically normal as $N\to\infty$ with parameters $\left(Np_r(a), \sqrt{(\sigma_r^2(a)+\sigma^2\left[p_r'(a)\right]^2)N}\right)$, where*

$$p_r(a)=\frac{a^r}{r!}e^{-a},\ \sigma_r(a)=p_r(a)\left[1-p_r(a)-p_r(a)\frac{(a-r)^2}{a}\right]. \tag{1}$$

PROOF. Let $\alpha=\nu/N$ and express α as

$$\alpha=a+\eta_\nu\frac{\sigma}{\sqrt{N}}. \tag{2}$$

By the hypothesis of the theorem, the random variable η_ν is asymptotically normal with parameters (0, 1). Further, let us write μ_r as

$$\mu_r=Np_r(\alpha)+\theta_r\sigma_r(\alpha)\sqrt{N}. \tag{3}$$

If the conditional distribution μ_r is considered for nonrandom α, then, as shown in Subsection 2.2, the distribution of the random variable θ_r is asymptotically normal with parameters (0, 1) as $N\to\infty$ and

$$0<\alpha_0\leqslant\alpha\leqslant\alpha_1<\infty \tag{4}$$

for all $\alpha_0>0$ and $\alpha_1<\infty$. Therefore, under condition (4), the random variable θ_r does not depend asymptotically on α. Since the probability of the event (4) tends to 1 because of the asymptotic normality of ν, the random variables η_ν and θ_r are asymptotically independent. Substituting (2) into (3) and expanding $p_r(\alpha)$ and $\sigma_r(\alpha)$ at the point $\alpha = a$, we have

$$\mu_r = N\left[p_r(a) + p_r(a)(\alpha - a) + O((\alpha - a)^2)\right] + \\ + \theta_r \sqrt{N}\left[\sigma_r(a) + O(\alpha - a)\right] = \\ = Np_r(a) + \left[p_r'(a)\sigma\eta_\nu + \sigma_r(a)\theta_r\right]\sqrt{N} + O\left(\eta_\nu^2\right) + O(\eta_\nu \theta_r)$$

or

$$\frac{\mu_r - Np_r(a)}{\sqrt{N\left[(p_r'(a))^2\sigma^2 + \sigma_r^2(a)\right]}} = \\ = \frac{p_r'(a)\sigma\eta_\nu + \sigma_r(a)\theta_r}{\sqrt{(p_r'(a))^2\sigma^2 + \sigma_r^2(a)}} + O\left(\frac{\eta_\nu^2}{\sqrt{N}}\right) + O\left(\frac{|\eta_\nu \theta_r|}{\sqrt{N}}\right). \quad (5)$$

The random variable $\eta = \dfrac{p_r'(a)\sigma\eta_\nu + \sigma_r(a)\theta_r}{\sqrt{(p_r'(a))^2\sigma^2 + \sigma_r^2(a)}}$ is asymptotically normal with parameters (0, 1), since η_ν and θ_r are asymptotically independent and normal with parameters (0, 1). By virtue of the asymptotic normality of η_ν and θ_r, the remainders $O\left(\eta_\nu^2/\sqrt{N}\right)$ and $O\left(\eta_\nu\theta_r/\sqrt{N}\right)$ in (5) converge in probability to zero. The theorem is proved.

§3. Limit Theorems in the Left- and the Right-Hand Domains

We proved in Subsection 2.3 that in the left-hand r-domain, i.e., as $\dfrac{n^r}{N^{r-1}r!} \to \lambda$, the random variable $\mu_r(n, N)$ is distributed in the limit according to the Poisson law:

$$\lim \mathbf{P}\{\mu_r(n, N) = k\} = \frac{\lambda^k}{k!} e^{-\lambda}. \tag{1}$$

Closely allied to this result is the following theorem for random ν:

Theorem 1. *If $r \geqslant 2$ is fixed and if the distribution function of the random variable $\nu^r/r!N^{r-1}$ converges weakly to $\Psi(x)$ as $N \to \infty$, then*

$$\lim_{N\to\infty} \mathbf{P}\{\mu_r = k\} = \int_0^\infty \frac{x^k}{k!} e^{-x} d\Psi(x). \tag{2}$$

PROOF. Denote by $\eta(x)$ a random variable with a Poisson distribution with parameter x. For every k, the probability $\mathbf{P}\{\eta(x) \leqslant k\}$ decreases as x increases. Taking advantage of this fact and the limit relation (1) for $\mu_r(n, N)$ in the case of a nonrandom number of particles, we obtain

$$\mathbf{P}\{\eta(x_{s+1}) \leqslant k\} \leqslant \varliminf_{N\to\infty} \mathbf{P}\left\{\mu_r \leqslant k \,\middle|\, x_s < \frac{\nu^r}{r!\,N^{r-1}} \leqslant x_{s+1}\right\} \leqslant$$

$$\leqslant \varlimsup_{N\to\infty} \mathbf{P}\left\{\mu_r \leqslant k \,\middle|\, x_s < \frac{\nu^r}{r!\,N^{r-1}} \leqslant x_{s+1}\right\} \leqslant \mathbf{P}\{\eta(x_s) \leqslant k\} \tag{3}$$

for $x_s < x_{s+1}$. Multiplying (3) by $\Psi(x_{s+1}) - \Psi(x_s)$ and summing over some partition $0 < x_1 < x_2 < \ldots < x_n < \infty$, we have, both on the left and on the right, approximating sums that converge to the same integral $\int_0^\infty \mathbf{P}\{\eta(x) \leqslant k\}\, d\Psi(x)$. This implies (2).

As we know (see Subsections 1.5 and 2.4), as $\frac{n^2}{2N} \to \lambda$, the random variables $\mu_0(n, N) - N + n$ and $(n - \mu_1(n, N))/2$ have the same limit distribution as $\mu_2(n, N)$ has. A similar result holds for random ν.

Theorem 2. *If* $\mathbf{P}\left\{\frac{\nu^2}{2N} \leqslant x\right\}$ *converges weakly to* $\Psi(x)$ *as* $N \to \infty$, *the random variables* $\mu_0+N+\nu$ *and* $(\nu-\mu_1)/2$ *have the same limit distribution as* μ_2.

In the right-hand r-domain, i.e., for n, $N \to \infty$ in such a way that

$$\frac{n}{N} - \ln N - r \ln \ln N = a + o(1),$$

for $\mu_r(n, N)$ with a nonrandom number of particles we have the limit theorem:

$$\lim \mathbf{P}\{\mu_r(n, N) = k\} = \frac{\lambda^k}{k!} e^{-\lambda}, \tag{4}$$

where $\lambda = \frac{1}{r!} e^{-a}$ (see Subsections 1.1, 2.3, and 2.4).

For random ν, this statement becomes

Theorem 3. *If* $\mathbf{P}\left\{\frac{\nu - N \ln N - rN \ln \ln N}{N} \leqslant x\right\}$ *converges weakly to* $\Psi(x)$ *as* $N \to \infty$, *then, for any* $r \geqslant 0$ *and* $k \geqslant 0$,

$$\lim_{N\to\infty} \mathbf{P}\{\mu_r = k\} = \int_0^\infty \frac{1}{k!} \left(\frac{1}{r!} e^{-x}\right)^k \exp\left\{-\frac{e^{-x}}{r!}\right\} d\Psi(x).$$

This theorem can be proved with the help of (4) in the same way as Theorem 1 is.

§4. Further Results. References

The main results of this chapter can be found in Ivchenko, Medvedev and Sevast'yanov [26].

Kolchin and Chistyakov [33] give the following multidimensional generalization of a scheme with a random number of particles. Let there be N cells partitioned into n groups. The ith group contains N_i cells with a random number ν_i of particles allocated in these cells. If $\nu_i = n_i$, then in the ith group these n_i particles are allocated in N_i cells with equal probabilities, independently both of each other and of allocations of particles in other groups of cells. We denote by μ_{0i} the number of empty cells in the ith group. The two limit theorems which follow are proved in [33].

Theorem 1. *If the distribution function of the vector* $(\nu_1/N_1, \ldots, \nu_m/N_m)$ *converges weakly to* $\Psi(x_1, \ldots, x_m)$ *as* $N_1, \ldots, N_m \to \infty$, *the probability distribution of the vector* $(\mu_{01}/N_1, \ldots, \mu_{0m}/N_m)$ *converges weakly to the probability distribution of the vector* $(e^{-\xi_1}, \ldots, e^{-\xi_m})$, *where*

$$\mathbf{P}\{\xi_1 \leqslant x_1, \ldots, \xi_m \leqslant x_m\} = \Psi(x_1, \ldots, x_m).$$

Theorem 2. *If the random vector* $(\nu_1, \ldots, \nu_m)$ *is asymptotically normal as* $N \to \infty$ *with mean vector* $(a_1 N, \ldots, a_m N)$, *where* $a_i > 0$ *for* $i = 1, \ldots, m$, *and covariance matrix* $\|b_{ij} N\|$ *and if the numbers of cells* $N_1, \ldots, N_m$ *in the groups are such that* $\frac{N_i}{N} \to \gamma_i^2 > 0$, *for* $i = 1, \ldots, m$, *then the vector* $(\mu_{01}, \ldots, \mu_{0m})$ *is asymptotically normal with mean vector* $\left(N_1 e^{-a_1}, \ldots, N_m e^{-a_m}\right)$ *and covariance matrix*

$$\left\| N\left(\delta_{ij} \gamma_i \gamma_j \sigma(a_i) \sigma(a_j) - b_{ij} \gamma_i^2 \gamma_j^2 e^{-a_i - a_j}\right) \right\|,$$

where $\delta_{ii} = 1$, $\delta_{ij} = 0$ *for* $i \neq j$, *and* $\sigma^2(\alpha) = e^{-\alpha}(1 - e^{-\alpha} - \alpha e^{-\alpha})$.

Chapter VII

ALLOCATION OF PARTICLES BY COMPLEXES

§1. The Statement of the Problem. Moments

In this chapter, we shall examine the following scheme of allocation of particles. Let there be N cells into which n' complexes of particles are thrown independently of each other, m particles in each complex. Thus, $n = n'm$ is the total number of particles in the n' complexes. The particles of each complex are allocated in the cells one each, all C_N^m possible allocations being equiprobable. We denote by $\mu_r'(n, N)$ the number of cells containing exactly r particles after the trial described above. Instead of $\mu_r'(n, N)$, we shall sometimes write briefly $\mu_r'(n)$ or μ_r'. In the case $m=1$, we have the standard scheme of equiprobable allocation discussed in Chapters 1 and 2. We shall call the scheme described at the beginning of this section the scheme *of allocation by complexes.* From now on we shall refer to the equiprobable scheme of allocation as *scheme I,* and the scheme of allocation by complexes as *scheme II.*

For the distribution law of the random variable $\mu_0'(n, N)$, formulas similar to (1.1.1) and (1.1.2) can be derived:

$$\mathbf{P}\{\mu_0'(n, N) = k\} = C_N^k \left[\frac{C_{N-k}^m}{C_N^m}\right]^{n'} \cdot \mathbf{P}\{\mu_0'(n, N-k)=0\}, \quad (1)$$

$$\mathbf{P}\{\mu_0'(n, N) = 0\} = (C_N^m)^{-n'} \sum_{l=0}^{N} C_N^l (-1)^l (C_{N-l}^m)^{n'}. \quad (2)$$

For proving (1) and (2), we denote by A_i the event that the ith cell in scheme II is empty and we denote by $\overline{A}_i$ the complementary event. Then

$$\mathbf{P}\{\mu_0(n, N) = k\} = \sum_{1 \leqslant i < \ldots < i_k \leqslant N} \mathbf{P}\{A_{i_1} \ldots A_{i_k} \overline{A}_{j_1} \ldots \overline{A}_{j_{N-k}}\}, \quad (3)$$

where $\{j_1, \ldots, j_{N-k}\} = \{1, \ldots, N\} \setminus \{i_1, \ldots, i_k\}$. In (3), all the terms of the summation are equal to each other and the number of them is C_N^k. By the theorem of product probabilities, we have:

$$\mathbf{P}\{A_{i_1} \ldots A_{i_k} \overline{A}_{j_1} \ldots \overline{A}_{j_{N-k}}\} = \\ = \mathbf{P}\{A_{i_1} \ldots A_{i_k}\} \cdot \mathbf{P}\{\overline{A}_{j_1} \ldots \overline{A}_{j_{N-k}} | A_{i_1} \ldots A_{i_k}\},$$

where

$$\mathbf{P}\{A_{i_1} \ldots A_{i_k}\} = \left(\frac{C_{N-k}^m}{C_N^m}\right)^{n'}$$

and

$$\mathbf{P}\{\overline{A}_{j_1} \ldots \overline{A}_{j_N} | A_{i_1} \ldots A_{i_k}\} = \mathbf{P}\{\mu_0'(n, N-k) = 0\}.$$

From this we obtain (1). To prove (2), we need to use the

formula for the probability of the union of events in

$$\mathbf{P}\{\mu_0'(n, N)>0\} = \mathbf{P}\left\{\bigcup_{i=1}^{N} A_i\right\}.$$

Arguing in the same way as in Subsection 2.1, we find the precise formula for the mathematical expectation μ_r'. Let us represent μ_r' as the sum $\mu_r' = \sum_{i=1}^{N} \theta_i$, where $\theta_i=1$ if there are exactly r particles in the ith cell and $\theta_i=0$ in all other cases. Since $\mathbf{M}\mu_r' = N\mathbf{P}\{\theta_i = 1\}$, we have:

$$\mathbf{M}\mu_r' = NC_{n'}^r\left(1-\frac{m}{N}\right)^{n'-r}\left(\frac{m}{N}\right)^r.$$

From this formula, results analogous to formulas (2.1.6) and (2.1.7) of Theorem 2.1.1 can be easily obtained:

Theorem 1. *For any n', N, r, and m we have the inequality*

$$\mathbf{M}\mu_r' \leqslant Np_r(\alpha)\, e^{rm/N}, \tag{4}$$

where $n=mn'$, $\alpha=\frac{n}{N}$, $p_r(\alpha)=\frac{\alpha^r}{r!}e^{-\alpha}$. *If* $\alpha=o(N)$ *as* $n,\ N\to\infty$, *then for fixed* r *and* m

$$\mathbf{M}\mu_r' \sim Np_r(\alpha). \tag{5}$$

In proving limit theorems for μ_0', we can use formulas (1) and (2). However for μ_0' and in general μ_r', there are no convenient generating functions such as those we used in scheme I. Hence, in investigating asymptotic properties of the random variables μ_r', we use a different method, comparing the corresponding distributions of μ_r' and μ_r obtained in scheme II and in some scheme I which closely "accompanies" it. Since the distributions of μ_r and μ_r' turn out to be close to each other, the same limit theorems hold true in scheme I as in scheme II.

§2. Limit Theorems

Let m particles be thrown independently into N cells. The probability q that every particle gets into a different cell from the others is equal to

$$q = \frac{N(N-1)\dots(N-m+1)}{N^m} > \left(1 - \frac{m}{N}\right)^m; \tag{1}$$

we denote the probability of the complementary event by $p = 1 - q$.

Assume that $n = n'm$ particles are allocated in N cells according to scheme I. Suppose that the particles occupy the cells by turns. We shall call the sequential groups containing m particles *complexes.* Therefore, if, according to scheme I, $n = n'm$ particles are thrown, we may suppose that m complexes of particles have been thrown independently. A complex in which the particles occupy m cells will be called *simple*; otherwise, the complex will be called *special.* By virtue of (1), the probability of the event that a given complex is simple is equal to q, and the probability of the event that complex is special is equal to p. Let us denote by ρ the random variable equal to the number of special complexes. It is seen that ρ has a binomial distribution:

$$\mathbf{P}\{\rho = k\} = C_{n'}^k p^k q^{n'-k}. \tag{2}$$

The construction given above will be used for proving the theorems in this section.

Theorem 1. *If* $\alpha = n/N \to 0$ *and* $m = o(1/\alpha)$ *as* $n, N \to \infty$, *then*

$$\mathbf{P}\{\mu_r \leqslant k\} - \mathbf{P}\{\mu_r' \leqslant k\} \to 0$$

uniformly with respect to k.

PROOF. It follows from (1) and (2) that

$$\mathbf{P}\{\rho=0\}=q^{n^*}>$$
$$>\left(1-\frac{m}{N}\right)^{mn'}=\left(1-\frac{m}{N}\right)^{n}\geqslant 1-\frac{mn}{N}=1+o(1) \qquad (3)$$

as $\alpha \to 0$. We write the probability $\mathbf{P}\{\mu_r \leqslant k\}$ as follows:

$$\mathbf{P}\{\mu_r \leqslant k\}=\mathbf{P}\{\rho=0\}\cdot\mathbf{P}\{\mu_r \leqslant k\,|\,\rho=0\}+\mathbf{P}\{\mu_r \leqslant k,\ \rho>0\}. \qquad (4)$$

The condition $\rho=0$ does not violate the independence of the complexes and implies that all the complexes are simple. Hence,

$$\mathbf{P}\{\mu_r \leqslant k\,|\,\rho=0\}=\mathbf{P}\{\mu'_r \leqslant k\}. \qquad (5)$$

From (4) and (5) we get

$$|\,\mathbf{P}\{\mu_r \leqslant k\}-\mathbf{P}\{\mu'_r \leqslant k\}\,|\leqslant \mathbf{P}\{\rho>0\},$$

from which the statement of the theorem follows by virtue of (3).

Corollary 1. It follows from Theorem 1 that, as $\alpha \to 0$, all Poisson limit theorems and normal limit theorems for μ_r are carried over to μ'_r.

Theorem 2. *If* $n, N \to \infty$ *in the central domain and if* $m = o(N^{1/4})$, *the random variable* μ'_r *is asymptotically normal with parameters* $(Np_r(\alpha),\ \sqrt{N\sigma_{rr}(\alpha)})$, *where* $\alpha=n/N$,

$$p_r(\alpha)=\frac{\alpha^r}{r!}e^{-\alpha},$$
$$\sigma_{rr}(\alpha)=p_r(\alpha)\left[1-p_r(\alpha)-p_r(\alpha)\frac{(\alpha-r)^2}{\alpha}\right].$$

PROOF. We consider scheme I the random variable $\mu_r(n, N)$ and again carry out the construction described at the beginning of this section. Let us denote by $A^{(l)}_{i_1\ldots i_l}$ the event that complexes numbered $i_1, \ldots, i_l$ are special and the remaining complexes are simple (in particular, $A^{(0)}=\{\rho=0\}$). From the formula of composite probability we have

$$\mathbf{P}\{\mu_r\leqslant k\}=$$
$$=\sum_{l=0}^{L}\sum_{1<i_1<\ldots i_l\leqslant N}\mathbf{P}\{\mu_r\leqslant k\mid A^{(l)}_{i_1\ldots i_l}\}\,\mathbf{P}\{A^{(l)}_{i_1\ldots i_l}\}+$$
$$+\mathbf{P}\{\mu_r\leqslant k,\ \rho>L\}. \quad (6)$$

The condition $A^{(l)}_{i_1\ldots i_l}$ preserves the independence of the complexes. For each $A^{(l)}_{i_1\ldots i_l}$, we perform the following additional trial: we replace all complexes numbered $i_1, i_2, \ldots, i_l$ by independent simple complexes obtained by scheme II. As a result, we have an allocation of n particles in N cells according to scheme II associated with the original allocation of particles according to scheme I. Let us consider in scheme II the random variable μ_r'. Since under the condition $A^{(l)}_{i_1\ldots i_l}$ the transition from μ_r to μ_r' involves variation of l complexes, this transition can lead to the variation of the content of no more than $2lm$ cells. Then, under the condition $A^{(l)}_{i_1\ldots i_l}$ we have the inequalities

$$\mu_r'-2lm\leqslant\mu_r\leqslant\mu_r'+2lm,$$

from which it follows that, for $l\leqslant L$,

$$\mathbf{P}\{\mu_r'\leqslant k-2Lm\mid A^{(l)}_{i_1\ldots i_l}\}\leqslant$$
$$\leqslant\mathbf{P}\{\mu_r\leqslant k\mid A^{(l)}_{i_1\ldots i_l}\}\leqslant\mathbf{P}\{\mu_r'\leqslant k+2Lm\mid A^{(l)}_{i_1\ldots i_l}\}. \quad (7)$$

From (6) and (7) it follows that

$$\mathbf{P}\{\mu_r\leqslant k\}\leqslant\sum_{l=0}^{L}\sum_{i_1<\ldots<i_l}\mathbf{P}\{\mu_r'\leqslant k+2Lm\mid A^{(l)}_{i_1\ldots i_l}\}\times$$
$$\times\mathbf{P}\{A^{(l)}_{i_1\ldots i_l}\}+\mathbf{P}\{\mu_r\leqslant k,\ \rho>L\},$$

$$\mathbf{P}\{\mu_r \leqslant k\} \geqslant \sum_{l=0}^{L} \sum_{i_1<\ldots<i_l} \mathbf{P}\{\mu_r' \leqslant k - 2Lm \mid A_{i_1\ldots i_l}^{(l)}\} \times$$
$$\times \mathbf{P}\{A_{i_1\ldots i_l}^{(l)}\} + \mathbf{P}\{\mu_r \leqslant k,\ \rho > L\}.$$

From this we obtain, for any k,

$$\mathbf{P}\{\mu_r \leqslant k\} \leqslant \mathbf{P}\{\mu_r' \leqslant k + 2Lm\} + \mathbf{P}\{\rho > L\}, \tag{8}$$

$$\mathbf{P}\{\mu_r \leqslant k\} \geqslant \mathbf{P}\{\mu_r' \leqslant k - 2Lm\} - 2\mathbf{P}\{\rho > L\}. \tag{9}$$

Let $L \to \infty$ in such a way that $\mathbf{P}\{\rho > L\} \to 0$ and $Lm = o(\sqrt{N})$. This can be done since in Chebyshev's inequality

$$\mathbf{P}\{\rho > L\} \leqslant \frac{\mathbf{M}\rho}{L} \tag{10}$$

the mathematical expectation

$$\mathbf{M}\rho = n'p < \frac{n}{m}\left(1 - \left(1 - \frac{m}{N}\right)^m\right) \leqslant \frac{mn}{N} = \alpha m$$

and $m = o(N^{1/4})$. Next, let $n, N \to \infty$ in such a way that, uniformly with respect to x,

$$\mathbf{P}\{\mu_r \leqslant Np_r + x\sqrt{N\sigma_{rr}}\} - G(x) \to 0,$$

where $G(x) = \frac{1}{\sqrt{2\pi}} \int_{-\infty}^{x} e^{-u^2/2}\, du.$ Setting $k - 2Lm = Np_r + x\sqrt{N\sigma_{rr}}$ in (9), we get

$$\mathbf{P}\{\mu_r' \leqslant Np_r + x\sqrt{N\sigma_{rr}}\} - G(x) \leqslant$$
$$\leqslant \mathbf{P}\{\mu_r \leqslant Np_r + x\sqrt{N\sigma_{rr}} + 2Lm\} - G(x) + 2\mathbf{P}\{\rho > L\} =$$
$$= \mathbf{P}\{\mu_r \leqslant Np_r + x'\sqrt{N\sigma_{rr}}\} - G(x') +$$
$$+ G(x') - G(x) + 2\mathbf{P}\{\rho > L\}, \tag{11}$$

where $x' = x + 2Lm/\sqrt{N\sigma_{rr}}$. From the restrictions imposed on L it follows that $x' - x \to 0$ and $\mathbf{P}\{\rho > L\} \to 0$. Therefore, the right-hand side of (11) approaches zero. Using (8), we estimate the difference $\mathbf{P}\{\mu_r' \leqslant Np_r + x\sqrt{N\sigma_{rr}}\} - G(x)$ from below and prove that this estimate also approaches zero.

The multidimensional limit theorem is proved in a similar way.

Theorem 3. *If $n, N \to \infty$ in the central domain $\left(\frac{n}{N} \to \alpha\right)$ and $m = o(N^{1/4})$, then, for any $0 \leqslant r_1 < r_2 < \ldots < r_s$, the distribution of the random vector*

$$\frac{\mu_{r_1}' - Np_{r_1}}{\sqrt{N}}, \ldots, \frac{\mu_{r_s}' - Np_{r_s}}{\sqrt{N}}$$

converges weakly to a multidimensional normal distribution with covariance matrix $\|\sigma_{r_i r_j}\|$,

$$\sigma_{rt} = p_r\left(\delta_{rt} - p_t - \frac{1}{\alpha}p_t(\alpha - r)(\alpha - t)\right),$$

$$p_r = \frac{\alpha^r}{r!}e^{-\alpha}.$$

We consider now the case $\alpha \to \infty$. First let us prove the following lemma.

Lemma 1. *If $n, N \to \infty$ in such a way that $\alpha = \frac{n}{N} \to \infty$, for the random variable $\mu' = \mu_0'(n) + \mu_1'(n) + \ldots + \mu_r'(n)$ we have*

$$\mathbf{P}\{\mu' < KNe^{-\alpha/2}\} \to 1 \tag{12}$$

for any $K > 0$.

PROOF. For any $0 \leqslant s \leqslant r$, we have by Chebyshev's inequality

$$\mathbf{P}\{\mu_s' \geqslant KNe^{-\alpha/2}\} \leqslant \frac{\mathbf{M}\mu_s'}{KNe^{-\alpha/2}}.$$

By virtue of (1.4), $\mathbf{M}\mu'_s \leqslant N \frac{\alpha^s}{s!} e^{-\alpha} e^{sm/N}$; hence, we have

$$\mathbf{P}\{\mu'_s > KNe^{-\alpha/2}\} \leqslant \frac{\alpha^s e^{-\alpha/2}}{Ks!} e^{sm/N} \to 0. \tag{13}$$

The assertion (12) follows from (13) and the inclusion

$$\{\mu' \geqslant x\} \subseteq \bigcup_{s=0}^{r} \left\{\mu'_s \geqslant \frac{x}{r+1}\right\}.$$

Theorem 4. *If* $\alpha \to \infty$ *as* n, $N \to \infty$, *then*

$$\mathbf{P}\{\mu'_r \leqslant k\} - \mathbf{P}\{\mu_r \leqslant k\} \to 0 \tag{14}$$

uniformly with respect to k.

PROOF. We consider first the case $\alpha = o(\sqrt{N})$. The start of the proof follows that of Theorem 2 including (6) and introducing the random variable μ'_r. But thereafter we proceed differently. Under the condition $A^{(l)}_{i_1 \ldots i_l}$, we introduce the random variable $\mu_r(n-lm)$, i.e., the number of cells containing exactly r particles not included in the complexes numbered $i_1, i_2, \ldots, i_l$. Next, if the missing complexes numbered $i_1, i_2, \ldots, i_l$ are allocated randomly in cells independently according to scheme II, we have the random variable $\mu'_r(n)$ related to the variable $\mu'_r(n-lm)$. If the missing complexes numbered $i_1, i_2, \ldots, i_l$ are allocated independently so that the occurrence of all special complexes is equiprobable and the occurrence of simple complexes is impossible, we arrive at the random variable $\mu_r(n)$ also related to $\mu'_r(n-lm)$. The distribution law of $\mu_r(n)$ is the same as that in scheme I. Let us prove that, with $l \leqslant L = [\alpha^2]$,

$$\mathbf{P}\{\mu'_r(n) = \mu'_r(n-lm) \mid A^{(l)}_{i_1 \ldots i_l}\} \to 1, \tag{15}$$

$$\mathbf{P}\{\mu_r(n) = \mu'_r(n-lm) \mid A^{(l)}_{i_1 \ldots i_l}\} \to 1. \tag{16}$$

In fact, in order to obtain the variable $\mu_r'(n)$ from $\mu_r'(n-lm)$, it is necessary to allocate l more complexes according to scheme II. If no new particles get into cells containing no more than r particles, then $\mu_r'(n-lm)=\mu_r'(n)$. Let $\mu'=\mu_0'(n-lm)+\ldots+\mu_r'(n-lm)$. By virtue of what was said above,

$$\mathbf{P}\{\mu_r'(n)=\mu_r'\ (n-lm)|A^{(l)}_{i_1\ldots i_l};\ \mu'\}\geqslant$$

$$\geqslant\left[\frac{(N-\mu')(N-\mu'-1)\ldots(N-\mu'-m+1)}{N(N-1)\ldots(N-m+1)}\right]^l\geqslant$$

$$\geqslant\left(1-\frac{\mu'}{N-m+1}\right)^{Lm}. \quad (17)$$

It follows from Lemma 1 that the right-hand side of (17), with probability tending to 1, is not smaller than the expression

$$\left(1-\frac{KNe^{-\alpha/2}}{N-m+1}\right)^{Lm},$$

which approaches 1 as $L=[\alpha^2]\to\infty$. Thus, (15) is proved. In proving (16), we write a similar inequality

$$\mathbf{P}\{\mu_r(n)=\mu_r'(n-lm)\,|A^l_{i_1,\ldots,i_l};\ \mu'\}\geqslant$$

$$\geqslant\left[\frac{(N-\mu')(N-\mu'-1)\ldots(N-\mu'-m+2)\,C_m^2}{N^m-N(N-1)\ldots(N-m+1)}\right]^l. \quad (18)$$

The expression on the right is the probability that newly thrown particles will miss the cells so far containing no more than r particles (the denominator $N^m-N(N-1)\ldots(N-m+1)$ is equal to the number of all special complexes, the numerator $(N-\mu')(N-\mu'-1)\ldots(N-\mu'-m+2)\,C_m^2$ is equal to the number of ways of allocating m particles in $m-1$ cells containing more than r particles). We estimate from below the right-hand side of (18) by the expression

$$\left[\frac{(N-\mu'-m+2)^{m-1}C_m^2}{N^{m-1}C_m^2+O(N^{m-2})}\right]^L \geqslant$$
$$\geqslant\left(1-\frac{\mu'+m-2}{N}\right)^{mL}\left(1+O\left(\frac{1}{N}\right)\right)^L. \quad (19)$$

The second factor in (19) tends to 1 since $L=[\alpha^2]=o(N)$, whence $L/N\to 0$. The first factor is estimated with the aid of Lemma 1 in the same way as in (17). From (10) with $L=[\alpha^2]$, we have $\mathbf{P}\{\rho>L\}\leqslant \mathbf{M}\rho/L=O(1/\alpha)\to 0$. Returning to (6) and using (15) and (16), we arrive at (14). The case of convergence of $\alpha\to\infty$ more rapidly than $o(\sqrt{N})$ leads us to the fact that

$$Np_r(\alpha)=N\frac{\alpha^r}{r!}e^{-\alpha}\to 0 \text{ and}$$
$$\mathbf{M}\mu_r\leqslant Np_r(\alpha)e^{r/N}\to 0, \qquad \mathbf{M}\mu_r'\leqslant Np_r(\alpha)e^{rm/N}\to 0.$$

By Chebyshev's inequality, we have, for any $k\geqslant 1$,

$$\mathbf{P}\{\mu_r>k\}\leqslant\frac{\mathbf{M}\mu_r}{k}\leqslant\mathbf{M}\mu_r\to 0, \qquad \mathbf{P}\{\mu_r'>k\}\leqslant\mathbf{M}\mu_r'\to 0,$$

from which (14) follows. The theorem is proved.

Corollary 2. It follows from Theorem 4 that as $\alpha\to\infty$ the Poisson theorem and the integral normal limit theorem for μ_r are carried over to μ_r'.

§3. Further Results. References

The scheme of allocation by complexes is mentioned in Markov [85]. In particular, formula (1.2) is given there. The method for investigating the asymptotic properties of the distribution laws of μ_r' via the "companion" schemes of allocation was first applied in Sevast'yanov [43].

In Pólya [90], Békéssy [60, 61], and Ivchenko and Medvedev

[25], a study is made of the random variable ν'_k equal to the number of complexes thrown according to scheme II that will cause at least k cells to be occupied for the first time. The asymptotic properties of the random variables $m\nu'_k$ and ν_k in scheme I (see Subsection 1.2) are analogous. For instance, as $k, N \to \infty$ in such a way that $\frac{k}{N} \to \gamma$, where $0<\gamma<1$,

$$\mathbf{M}\nu'_k \sim \frac{N}{m} \ln \frac{1}{1-\gamma}, \tag{1}$$

$$\mathbf{D}\nu'_k \sim \frac{N}{m^2}\left(\frac{\gamma}{1-\gamma} - \ln \frac{1}{1-\gamma}\right) \tag{2}$$

(compare with Theorem 1.2.1). Under the same conditions, ν'_k is asymptotically normal with parameters $\left(\mathbf{M}\nu'_k, \sqrt{\mathbf{D}\nu'_k}\right)$. If $|N-k| \leqslant C < \infty$, the random variable $m\nu'_k$ has asymptotically (as $N,\ k \to \infty$) the same distribution as ν_k has in Theorem 1.2.4. If $\frac{k^2}{2N} \to \lambda$, then $\nu'_k + \left[-\frac{k}{m}\right]$ has a discrete limit distribution $\{p_a\}$ in which the probabilities are determined by the equations

$$p_0 = e^{-\lambda} \sum_{i=0}^{m-s} \frac{\lambda^i}{i!}, \qquad p_a = e^{-\lambda} \sum_{i=am-s+1}^{(a+1)m-s} \frac{\lambda^i}{i!}, \qquad a \geqslant 1,$$

for $k \equiv s \pmod m$, $1 \leqslant s \leqslant m$.

Chapter VIII

GENERALIZATION OF THE PROBLEM OF ALLOCATION AND CYCLES OF RANDOM PERMUTATIONS

§1. A Generalized Scheme of Allocation of Particles

In the schemes of allocation of particles to cells investigated in the previous chapters, the cells could be filled by means of sequential allocations of particles to cells with simple probability relations between the sequential allocations. If this is not the case, any system of N nonnegative-integer-valued random variables $\eta_1, \ldots, \eta_N$ such that $\eta_1+\ldots+\eta_N=n$ can be treated as a generalized scheme of allocation of n particles into N cells and η_i can be interpreted as the number of particles in a cell numbered i, for $i=1, \ldots, N$.

In some probabilistic problems of combinatorics, generalized schemes of allocation are employed in which the joint distribution of $\eta_1, \ldots, \eta_N$ is representable as:

$$\mathbf{P}\{\eta_i=k_i,\ i=1, \ldots, N\}= \\ =\mathbf{P}\{\xi_i=k_i,\ i=1,\ldots, N \mid \xi_1+\ldots+\xi_N=n\}, \quad (1)$$

where $\xi_1, \ldots, \xi_N$ are independent identically distributed non-negative-integer-valued random variables. This generalized scheme of allocation of particles into cells is determined by the distribution of the random variable ξ_1. If we take for the distribution of ξ_1 a Poisson law, (1) becomes a polynomial distribution and then the generalized scheme coincides with the equiprobable scheme of allocation of particles.

Let us denote by $\mu_r(n, N)$ the number of cells containing r particles in the generalized scheme of allocation with the distribution (1). Then

$$\mathbf{P}\{\mu_r(n, N) = k\} = C_N^k p_r^k (1 - p_r)^{N-k} \times \\ \times \frac{\mathbf{P}\{\xi_1^{(r)} + \ldots + \xi_{N-k}^{(r)} = n - kr\}}{\mathbf{P}\{\xi_1 + \ldots + \xi_N = n\}}, \quad (2)$$

where $p_r = \mathbf{P}\{\xi_1 = r\}$, $\mathbf{P}\{\xi_1^{(r)} = k\} = \mathbf{P}\{\xi_1 = k \mid \xi_1 \neq r\}$ and the random variables $\xi_1^{(r)}, \ldots, \xi_N^{(r)}$ are independent and identically distributed. The proof of (2) is the same as that of Lemma 2.3.1. Therefore, proving limit theorems for $\mu_r(n, N)$ reduces in the generalized scheme to investigating the limit behavior of the sums of independent terms.

In the generalized schemes of allocation of particles, a rather simple approach to the investigation of the terms of the order sequence $\eta_{(1)}, \ldots, \eta_{(N)}$ obtained by arranging $\eta_1, \ldots, \eta_N$ in nondecreasing order is possible.

Let $\xi_1^{(A)}, \ldots, \xi_N^{(A)}$ denote independent identically distributed random variables for which

$$\mathbf{P}\{\xi_i^{(A)} = k\} = \mathbf{P}\{\xi_i = k \mid \xi_i \notin A\}, \quad (3)$$

where A is a subset of the set of natural numbers, and $\mathbf{P}\{\xi_i \notin A\} > 0$. Let us define $\zeta_N = \xi_1 + \ldots + \xi_N$ and $\zeta_N^{(A)} = \xi_1^{(A)} + \ldots + \xi_N^{(A)}$.

The following two lemmas reduce the investigation of random variables $\mu_r(n, N)$ and the terms of the order sequence

$\eta_{(1)}, \ldots, \eta_{(N)}$ to investigation of the limit behavior of the sums of independent terms.

Lemma 1. *We have*

$$\mathbf{P}\{\mu_{r_1}(n, N) = k_1, \ldots, \mu_{r_s}(n, N) = k_s\} = \frac{N! p_{r_1}^{k_1} \ldots p_{r_s}^{k_s} (1 - p_{r_1} - \ldots - p_{r_s})^{n - k_1 r_1 - \ldots - k_s r_s}}{k_1! \ldots k_s! (N - k_1 - \ldots - k_s)!} \times$$
$$\times \frac{\mathbf{P}\left\{\zeta_{N-k_1 \ldots -k_s}^{(r_1, \ldots, r_s)} = n - k_1 r_1 - \ldots - k_s r_s\right\}}{\mathbf{P}\{\zeta_N = n\}}, \tag{4}$$

where $s-1$, $k_1, \ldots, k_s$, $r_1, \ldots, r_s$ *are nonnegative integers and* $r_1, \ldots, r_s$ *are distinct.*

Equation (4) is a simple generalization of (2). Both are proved in a similar way.

Lemma 2. *We have*

$$\mathbf{P}\{\eta_{(m)} \leqslant r\} = 1 - \sum_{l=0}^{m-1} C_N^l (\mathbf{P}\{\xi_1 \leqslant r\})^l \times$$
$$\times (\mathbf{P}\{\xi_1 > r\})^{N-l} \frac{\mathbf{P}\left\{\zeta_l^{(\bar{A}_r)} + \zeta_{N-l}^{(A_r)} = n\right\}}{\mathbf{P}\{\zeta_N = n\}}, \tag{5}$$

$$\mathbf{P}\{\eta_{(N-m+1)} \leqslant r\} = \sum_{l=0}^{m-1} C_N^l (\mathbf{P}\{\xi_1 > r\})^l \times$$
$$\times (\mathbf{P}\{\xi_1 \leqslant r\})^{N-l} \frac{\mathbf{P}\left\{\zeta_l^{(A_r)} + \zeta_{N-l}^{(\bar{A}_r)} = n\right\}}{\mathbf{P}\{\xi_N = n\}}, \tag{6}$$

where A_r *is the set of nonnegative integers not exceeding* r *and* $\bar{A}_r$ *is the complement of* A_r *in the set of nonnegative integers.*

The proof of Lemma 2 is similar to that of Lemma 2.6.1.

In the future, we shall need a generalized scheme of allocation of particles with a random number of cells. Let ν_n be a nonnegative-integer-valued random variable. The allocation of n

particles to ν_n cells is described by the joint distribution of the variables $\eta_1, \eta_2, \ldots, \eta_{\nu_n}$, where η_i is interpreted as the number of particles in the ith cell for $i=1, \ldots, \nu_n$. The joint distribution of $\eta_1, \ldots, \eta_{\nu_n}$ can be given by the distribution of the random variable ν_n and the conditional distributions of $\eta_1, \ldots, \eta_N$ under the condition $\nu_n = N$. We assume that for each n, N there exist independent identially distributed integer-valued random variables $\xi_1, \ldots, \xi_N$ such that

$$\mathbf{P}\{\eta_i = k_i,\ i=1, \ldots, N \mid \nu_n = N\} = \\ = \mathbf{P}\{\xi_i = k_i,\ i=1, \ldots, N \mid \xi_1 + \ldots + \xi_N = n\}. \quad (7)$$

Therefore, the generalized scheme of allocation of particles with a random number of cells is given by the distributions of the random variables ν_n and ξ_1.

Let us denote by $\mu_r(n, \nu_n)$ the number of cells containing exactly r particles in the generalized scheme of allocation with the random number of cells ν_n given by the distribution (7). Let us arrange the occupancies of cells $\eta_1, \ldots, \eta_{\nu_n}$ in non-decreasing order. Let us also denote by $\eta^*_{(1)}, \ldots, \eta^*_{(\nu_n)}$ the order sequence thus obtained. We define the random variables s_m by: $s_m = \eta^*_{(m)}$ if $\nu_n \geqslant m$, and $s_m = 0$ if $\nu_n < m$. It can easily be seen that the conditional distribution of $\mu_r(n, \nu_n)$ under the condition $\nu_n = N$ coincides with the distribution of $\mu_r(n, N)$; the conditional distribution of s_m under the condition $\nu_n = N$, $N \geqslant m$, coincides with the distribution of $\eta_{(m)}$ in the generalized scheme of allocation of particles given by the distribution in (1).

§2. The Generalized Scheme of Allocation of Particles and Random Permutations

In this section, we study how random permutations are related to the generalized scheme of allocation of particles with a random number of cells.

Let us consider a one-to-one mapping T of the set $\{1, 2, \ldots, n\}$ onto itself. This mapping can be written

$$T = \begin{pmatrix} 1 & 2 & \ldots & n \\ i_1 & i_2 & \ldots & i_n \end{pmatrix}, \tag{1}$$

where i_k denotes the image of k under the mapping T. We call the mapping T or the scheme (1) a permutation of degree n. The number of permutations of degree n is equal to $n!$. The mapping T can be represented as an oriented graph G with n vertices (numbered $1, 2, \ldots, n$) and n edges. The vertices numbered k and l are connected by the edge directed from k to l if and only if T maps the element k into l (i.e., $i_k = l$). Only one edge starts from each vertex of the graph G, and, since the mapping is one-to-one, exactly one edge ends at each vertex. Hence, the graph G consists of separate connected components each of which is a cycle. We study in this section the characteristics associated with the permutation cycles.

Let us consider a set of permutations of degree n, and, for each permutation in this set, let us consider the vector $a = (\alpha_1, \ldots, \alpha_n)$, where α_r is the number of cycles of length r in that permutation. Denote by $\varkappa_n$ the total number, $\alpha_1 + \ldots + \alpha_n$, of cycles in the permutation. We arrange the cycles in nondecreasing order of length. We denote by S_m the length of the mth cycle in the sequence thus obtained. (We take $S_m = 0$ if $\varkappa_n < m$).

On the set of permutations of degree n, let there be given a probability distribution with each permutation assigned the probability $1/n!$. Then the components $\alpha_1, \ldots, \alpha_n$ of the vector a, the number $\varkappa_n$ of cycles, and the ordered lengths S_m of the cycles will be random variables.

Let us investigate the limit behavior of the distributions of these random variables as $n \to \infty$. First we find the precise expressions for the joint distribution of $\alpha_1, \ldots, \alpha_n$ and for the distribution of $\varkappa_n$ cycles.

Theorem 1. *If $m_1, \ldots, m_n$ are nonnegative integers, then*

$$\mathbf{P}\{\alpha_i=m_i,\ i=1,\ldots,n\}=\begin{cases}\prod_{r=1}^{n}\frac{1}{r^{m_r}m_r!} & \text{if } \sum_{r=1}^{n} rm_r=n,\\ 0 & \text{otherwise.}\end{cases}$$

PROOF. Since each permutation has the same probability, for calculating the probability $\mathbf{P}\{\alpha_i=m_i, i=1,\ldots, n\}$ it suffices to find the number $a(m_1, \ldots, m_n)$ of distinct permutations having m_i cycles of length i for $i=1, \ldots, n$. Let us choose one permutation with such a cyclic composition, and let us place n elements in the vertices of the corresponding graph G in all $n!$ possible ways. In this case, all permutations with cyclic composition $(m_1, \ldots, m_n)$ show up, and each permutation occurs $b(m_1, \ldots, m_n)$ times. Let us find this number. The permutations resulting from interchanging the elements may coincide in two ways: first, because the cycles including the same elements in the same order do not differ among themselves; second, because the relative arrangement of the cycles in the permutation is of no importance. The first reason leads to the fact that each cycle of length r generates r identical permutations and therefore each permutation occurs $1^{m_1}2^{m_2}\ldots n^{m_n}$ times. The second reason leads to the fact that m_r cycles of length r generate $m_r!$ identical permutations and therefore each permutation occurs $m_1!\, m_2! \ldots m_n!$ times. Therefore, if the nonnegative integers $m_1, \ldots, m_n$ are such that $m_1+2m_2+\ldots+nm_n=n$, there exists at least one permutation with cyclic composition $(m_1, \ldots, m_n)$. In this case,

$$b(m_1, \ldots, m_n) = 1^{m_1}2^{m_2}\ldots n^{m_n}m_1!\ldots! m_n!,$$
$$a(m_1, \ldots, m_n) = \frac{n!}{b(m_1, \ldots, m_n)};$$

if $m_1+2m_2+\ldots+nm_n\neq n$, then $a(m_1, \ldots, m_n)=0$. This implies the statement of the theorem, since, by the classical definition of probability,

$$\mathbf{P}\{\alpha_i = m_i,\ i = 1, \ldots, n\} = \frac{a(m_1, \ldots, m_n)}{n!}.$$

Let us find the distribution of the number $\varkappa_n$ of cycles in a random permutation.

Theorem 2. *We have*

$$\mathbf{P}\{\varkappa_n = N\} = \frac{1}{N!} \sum_K \frac{1}{k_1 \ldots k_N}, \qquad N = 1, \ldots, n,$$

where the summation is over the set of integers

$$K = \{k_i \geqslant 1,\ i = 1, \ldots, N;\ k_1 + \ldots + k_N = n\}.$$

Proof. We introduce the generating function

$$A_n(t_1, \ldots, t_n) =$$

$$= \sum_{m_1, \ldots, m_n = 0}^{\infty} \mathbf{P}\{\alpha_i = m_i,\ i = 1, \ldots, n\}\ t_1^{m_1} \ldots t_n^{m_n} =$$

$$= \sum_{M_n} \frac{1}{m_1! \ldots m_n!} \left(\frac{t_1}{1}\right)^{m_1} \left(\frac{t_2}{2}\right)^{m_2} \ldots \left(\frac{t_n}{n}\right)^{m_n},$$

where the summation is over the set of integers $M_n = \{m_i \geqslant 0,\ i = 1, \ldots, n;\ m_1 + 2m_2 + \ldots + nm_n = n\}$. Let $A_0 = 1$. It is easily seen that $A_n(t_1, \ldots, t_n)$ is the coefficient of u^n in the expansion of $\exp\left(ut_1 + \frac{u^2 t_2}{2} + \ldots\right)$ in powers of u:

$$\sum_{n=0}^{\infty} A_n(t_1, \ldots, t_n) u^n = \exp\left(\sum_{n=1}^{\infty} \frac{u^n t_n}{n}\right). \tag{2}$$

The probability $\mathbf{P}\{\varkappa_n = N\}$ is the coefficient of t^N in the expansion of $A_n(t, \ldots, t)$ in powers of t and hence is the coefficient of $u^n t^N$ in $\sum_{n=0}^{\infty} A_n(t, \ldots, t)\, u^n$. Using (2), we find that

$$\sum_{n=0}^{\infty} A_n(t, \ldots, t)\, u^n = e^{t\left(u+\frac{u^2}{2}+\frac{u^3}{3}+\ldots\right)}.$$

The coefficients of t^N is equal to $\frac{1}{N!}\left(u+\frac{u^2}{2}+\frac{u^3}{3}+\ldots\right)^N$, and the coefficient of u^n is equal to $\frac{1}{N!}\sum_K \frac{1}{k_1 \ldots k_N}$. The theorem is proved.

We consider next the generalized scheme of allocation of particles with the random number of cells given by the distribution in (1.7), in which

$$\mathbf{P}\{\nu_n = N\} = \mathbf{P}\{\varkappa_n = N\}, \quad N = 1, \ldots, n, \tag{3}$$

$$\mathbf{P}\{\xi_1 = k\} = -\frac{\theta^k}{k \ln(1-\theta)}, \quad 0<\theta<1, \ k = 1, 2, \ldots. \tag{4}$$

Lemma 1. *The distribution of the number $\varkappa_n$ of cycles in a random permutation of degree n can be represented as*

$$\mathbf{P}\{\varkappa_n = N\} = \frac{(-\ln(1-\theta))^N}{N!\theta^n}\,\mathbf{P}\{\xi_1 + \ldots + \xi_N = n\},$$
$$N = 1, 2, \ldots, n. \tag{5}$$

Proof. It is easily seen that, for independent uniformly distributed random variables with the distribution (4),

$$\mathbf{P}\{\xi_1 + \ldots + \xi_N = n\} = \frac{\theta^n}{(-\ln(1-\theta))^N}\sum_K \frac{1}{k_1 \ldots k_N},$$

where $K = \{k_i \geqslant 1,\ i=1, \ldots, N;\ k_1+\ldots+k_N=n\}$. From this the statement of the theorem follows by virtue of Theorem 2.

Using Lemma 1, it is easy to establish that in the generalized scheme of allocation of particles given by the distributions in (3) and (4),

$$\mathbf{P}\{\eta_i=k_i,\ i=1,\dots,\ N\,|\,\nu_n=N\}=$$

$$=\begin{cases}\dfrac{\theta^n}{(-\ln(1-\theta))^N k_1\dots k_N\mathbf{P}\{\xi_1+\dots+\xi_N=n\}} & \text{if } \sum_{i=1}^{N} k_i=n,\\ 0 & \text{otherwise.}\end{cases} \tag{6}$$

The relation between the distribution of the vector $a=(\alpha_1,\dots,\ \alpha_n)$ describing the cyclic structure of a random permutation and the distribution of the random variables $\mu_1(n,\ \nu_n),\ \dots,\ \mu_n(n,\nu_n)$ in the generalized scheme of allocation of particles with a random number of cells given by the distributions (3) and (4) plays an essential role.

Theorem 3. *We have*

$$\mathbf{P}\{\alpha_i=m_i,\ i=1,\dots,\ n\}=\mathbf{P}\{\mu_i(n,\ \nu_n)=m_i,\ i=1,\dots,\ n\}.$$

PROOF. Using (6), we easily obtain

$$\mathbf{P}\{\mu_i(n,N)=m_i,\ i=1,\dots,n\}=$$

$$=\begin{cases}\dfrac{N!\theta^n\prod_{i=1}^{n}\dfrac{1}{i^{m_i}m_i!}}{(-\ln(1-\theta))^N\mathbf{P}\{\xi_1+\dots+\xi_N=n\}} & \text{if } \sum_{i=1}^{n} m_i=N \\ & \text{and } \sum_{i=1}^{n} im_i=n,\\ 0 & \text{otherwise.}\end{cases} \tag{7}$$

By the formula of composite probability,

$$\mathbf{P}\{\mu_i(n,\nu_n)=m_i,\ i=1,\dots,n\}=$$

$$=\sum_{N=1}^{\infty}\mathbf{P}\{\nu_n=N\}\mathbf{P}\{\mu_i(n,N)=m_i,\ i=1,\dots,n\}=$$

$$=\mathbf{P}\{\nu_n=m\}\mathbf{P}\{\mu_i(n,m)=m_i,\ i=1,\dots,n\},$$

where $m=m_1+\ldots+m_n$. Hence by virtue of (5), (3), and (7)

$$\mathbf{P}\{\mu_i(n, \nu_n)=m_i,\ i=1,\ldots,n\}=$$

$$=\begin{cases}\prod_{i=1}^{n}\dfrac{1}{i^{m_i}m_i!} & \text{if } \sum_{i=1}^{n} im_i=n,\\ 0 & \text{otherwise.}\end{cases}$$

The theorem is proved.

Since the random variables s_m defined in Section 1 can be expressed in terms of $\mu_r(n, \nu_n)$, it follows by Theorem 3 that distributions s_m coincide with the distributions of the random variables S_m equal to the length of the mth cycle in a random permutation of degree m. Therefore, by means of Theorem 3 investigation of $\alpha_1, \ldots, \alpha_n$ and the related variables S_m reduces to investigation of the generalized scheme (given by the distributions (3) and (4)) of allocation of particles with a random number of cells. We shall study this scheme as follows. We first consider the scheme under the condition that ν_n assumes a fixed value N and then average the results with respect to the distribution of ν_n.

By means of Lemmas 1.1 and 1.2, the investigation of $\mu_r(n, N)$ and the terms of an order sequence $\eta_{(1)}, \ldots, \eta_{(N)}$ in the generalized scheme of allocation of n particles to N cells reduces to investigation of the sums of independent identically distributed terms $\zeta_N=\xi_1+\ldots+\xi_N$ and $\zeta_N^{(A)}=\xi_1^{(A)}+\ldots+\xi_N^{(A)}$. The auxiliary results relevant to the asymptotic behavior of these sums for random variables with the distribution given by (4) can be found in the next section.

§3. Some Properties of a Logarithmic Series Distribution

In this section, the local limit theorems are given for the sums $\zeta_N=\xi_1+\ldots+\xi_N$ and sums of the form $\zeta_N^{(A)}=\xi_1^{(A)}+\ldots+\xi_N^{(A)}$ consisting of independent uniformly distributed random variables. In this case, ξ_1 has a logarithmic series distribution with parameter θ:

$$\mathbf{P}\{\xi_1 = k\} = -\frac{\theta^k}{k \ln(1-\theta)}, \qquad k = 1, 2, \ldots, \tag{1}$$

and the distribution of $\xi_1^{(A)}$ is connected with (1) by (1.3). The theorems on ζ_N and $\zeta_N^{(A)}$ are of an auxiliary nature and used in Section 4 for proving the limit theorems on cycles of random permutations. A detailed proof of the limit theorem is given for the sum ζ_N. The proof of the theorems for sums of the form $\zeta_N^{(A)}$ is essentially the same.

Throughout what follows, the parameter θ of the distribution (1) is chosen to be $\theta = 1 - n^{-1}$ because in this case the mathematical expectation

$$\mathbf{M}\xi_1 = -\frac{\theta}{(1-\theta)\ln(1-\theta)} = \frac{n-1}{\ln n}.$$

It will follow from subsequent results that the mean number of cycles of a random permutation is $\ln n$. Hence the value of $\mathbf{M}\xi_1$ for $\theta = 1 - n^{-1}$ is equivalent to the average length of the cycle of a random permutation.

Let us obtain some estimates for the characteristic function of a logarithmic series distribution with the parameter $\theta = 1 - n^{-1}$, this function being equal to

$$\varphi_n(t) = -\frac{1}{\ln n} \ln\left(1 - e^{it} + \frac{1}{n} e^{it}\right).$$

We write $\varphi_n(t)$ as

$$\varphi_n(t) = -\frac{1}{\ln n}\left[\ln\left(\frac{1}{n} - it\right) + \ln\left(1 + \Psi_1(t) + \Psi_2(t)\right)\right], \tag{2}$$

where

$$\Psi_1(t) = \frac{1+it-e^{it}}{\frac{1}{n}-it}, \qquad \Psi_2(t) = \frac{e^{it}-1}{n\left(\frac{1}{n}-it\right)}.$$

For $\Psi_1(t)$ and $\Psi_2(t)$, we have the estimates

$$|\Psi_1(t)| \leqslant \frac{|e^{it}-1-it|}{|t|} \leqslant \frac{|t|}{2}, \tag{3}$$

$$|\Psi_2(t)| \leqslant \frac{|e^{it}-1|}{n|t|} \leqslant \frac{1}{n}. \tag{4}$$

Using the explicit form of $\varphi_n(t)$, the representation (2), and the estimates (3) and (4), we easily obtain the following properties of $\varphi_n(t)$.

Lemma 1. *If* $N=\ln n+o(\ln n)$ *as* $n\to\infty$, *then for any fixed* t,

$$\varphi_n^N\left(\frac{t}{n}\right) = \frac{1}{1-it} + o(1).$$

Lemma 2. *If* $N=\ln n+o(\ln n)$ *as* $n\to\infty$, *then for any* $\delta>0$ *there exist positive numbers* ε *and* c *such that the inequality*

$$\left|\varphi_n^N\left(\frac{t}{n}\right)\right| < \frac{c}{(1+|t|)^{1-\delta}}$$

is satisfied for $|t/n| \leqslant \varepsilon$.

Lemma 3. *If* $N=\ln n+o(\ln n)$ *as* $n\to\infty$, *then for* $\pi \geqslant |t| \geqslant \varepsilon > 0$, *we have, starting with some* n,

$$|\varphi_n(t)| < \frac{1}{3}.$$

Lemma 4. *If* $n\to\infty$, *there exists* $\varepsilon>0$ *such that, for* $|t/n|<\varepsilon$,

$$\left|\frac{1}{n}\varphi_n'\left(\frac{t}{n}\right)\right| < \frac{c}{(1+|t|)\ln n},$$

where c *is some constant.*

It follows from Lemma 1 that, if $N = \ln n + o(\ln n)$ as $n \to \infty$, then the distributions of the normalized sums ζ_N/n converge to an exponential distribution. In fact, there is a local approximation for these distributions.

Theorem 1. *If*

$$N = \ln n + o(\ln n), \quad z = k/n,$$

as $n \to \infty$, where k are natural numbers, then uniformly with respect to $z \geqslant z_0 > 0$

$$n\mathbf{P}\left\{\frac{1}{n}\zeta_N = z\right\} = e^{-z} + o(1).$$

PROOF. By the inversion formula, the probability

$$\mathbf{P}\{\zeta_N = k\} = \mathbf{P}\left\{\frac{1}{n}\zeta_N = z\right\}$$

is representable in the form:

$$\mathbf{P}\left\{\frac{1}{n}\zeta_N = z\right\} = \frac{1}{2\pi n}\int_{-\pi n}^{\pi n} e^{-itz}\varphi_n^N\left(\frac{t}{n}\right)dt;$$

moreover

$$e^{-z} = \frac{1}{2\pi}\int_{-\infty}^{\infty}\frac{e^{-itz}}{1-it}dt.$$

Hence

$$2\pi n\mathbf{P}\left\{\frac{1}{n}\zeta_N = z\right\} - 2\pi e^{-z} = I_1 + I_2 + I_3 + I_4,$$

where

$$I_1 = \int_{-A}^{A} e^{-itz}\left[\varphi_n^N\left(\frac{t}{n}\right) - \frac{1}{1-it}\right] dt\,,$$

$$I_2 = \int_{A\leqslant|t|\leqslant\varepsilon n} e^{-itz}\,\varphi_n^N\left(\frac{t}{n}\right) dt\,,$$

$$I_3 = \int_{\varepsilon n<|t|\leqslant\pi n} e^{-itz}\,\varphi_n^N\left(\frac{t}{n}\right) dt\,,$$

$$I_4 = \int_{A\leqslant|t|} \frac{e^{-itz}}{1-it}\, dt\,,$$

the constants A and ε to be chosen later. By Lemma 1, the integral I_1 approaches zero for any fixed A. By Lemma 3, we have $|I_3| < \pi n\left(\frac{1}{3}\right)^{\ln n + o(\ln n)} \to 0$. To estimate the integrals I_2 and I_4, we integrate by parts. For I_4, this leads to

$$\int_A^\infty \frac{e^{-itz}}{1-it}\, dt = -\left.\frac{e^{-itz}}{iz\,(1-it)}\right|_A^\infty + \frac{1}{z}\int_A^\infty \frac{e^{-itz}}{(1-it)^2}\, dt.$$

Hence

$$|I_4| \leqslant \frac{2}{z\sqrt{1+A^2}} + \frac{2}{z}\int_A^\infty \frac{dt}{(1+t)^2},$$

and, by choosing A appropriately, we can make $|I_4|$ arbitrarily small.

Similarly,

$$\int_A^{\varepsilon n} e^{-itz}\,\varphi_n^N\left(\frac{t}{n}\right) dt =$$

$$= -\frac{e^{-itz}}{iz}\,\varphi_n^N\left(\frac{t}{n}\right)\Bigg|_A^{\varepsilon n} + \frac{N}{iz}\int_A^{\varepsilon n} e^{-itz}\varphi_n^{N-1}\left(\frac{t}{n}\right)\frac{1}{n}\,\varphi_n'\left(\frac{t}{n}\right) dt.$$

Using the estimates given in Lemmas 2 and 4, we obtain

$$|I_2| \leqslant \frac{2}{z}\left|\varphi_n^N\left(\frac{A}{n}\right)\right| + \frac{2}{z}|\varphi_n^N(\varepsilon)| + \frac{c}{z}\int_A^\infty \frac{dt}{t^{2-\delta}},$$

and by choosing A suitably, we can make $|I_2|$ arbitrarily small since, by Lemmas 1 and 3,

$$\left|\varphi_n^N\left(\frac{A}{n}\right)\right| < \frac{c}{A}, \quad |\varphi_n^N(\varepsilon)| \leqslant \frac{1}{3^N}$$

and, by Lemma 1, we can take δ to be less than one in the integral in the right side of the inequality for $|I_2|$, thus proving the theorem.

The local limit theorem for the sum $\zeta_N^{(r_1,\ldots,r_s)}$, where $r_1,\ldots, r_s$ are fixed natural numbers, is proved in exactly the same way.

Theorem 2. *If* $N=\ln n+o(\ln n)$ *as* $n\to\infty$ *and* $z=k/n$, *where* k *is a natural number, then*

$$n\mathbf{P}\left\{\zeta_N^{(r_1,\ldots,r_s)}=z\right\}=e^{-z}+o(1)$$

uniformly with respect to $z\geqslant z_0>0$.

Next we consider the sums $\zeta_l^{(\bar{A}_r)}+\zeta_{N-l}^{(A_r)}$ appearing in Lemma 1.1, where A_r is the set of natural numbers not exceeding r and $\bar{A}_r$ is the complement of A_r in the set of all natural numbers and where the random variable ξ_1 has the distribution given by (1).

Let us denote by $\varphi_{(r)}(t)$ the characteristic function $\xi_1^{(A_r)}$ equal to

$$\varphi_{(r)}(t)=\left(\varphi_n(t)-\sum_{k=1}^{r} p_k e^{itk}\right)\left(1-\sum_{k=1}^{r} p_k\right)^{-1}.$$

We discuss first the case in which

$$r=\exp\{\alpha\ln n+t\sqrt{\ln n}\},\ 0<\alpha<1.$$

Lemma 5. *If* $r=\exp\{\alpha \ln n+t\sqrt{\ln n}\}$ *as* $n\to\infty$, *where* $0<\alpha<1$, *then if* $t=o(\sqrt{\ln n})$,

$$\mathbf{P}\{\xi_1\leqslant r\}=\alpha+\frac{t}{\sqrt{\ln n}}+O\left(\frac{1}{\ln n}\right).$$

PROOF. Since $\left(1-\frac{1}{n}\right)^k=1+O\left(\frac{k}{n}\right)$ for $k\leqslant r$, we have

$$\mathbf{P}\{\xi_1\leqslant r\}=\sum_{k=1}^{r}p_k=\sum_{k=1}^{r}\frac{\left(1-\frac{1}{n}\right)^k}{k\ln n}=\frac{1}{\ln n}\sum_{k=1}^{r}\frac{1}{k}+O\left(\frac{r}{n\ln n}\right).$$

It remains only to note that $\sum_{k=1}^{r}\frac{1}{k}=\ln r+O(1)$ as $r\to\infty$.

Lemma 6. *If* $r=\exp\{\alpha\ln n+o(\ln n)\}$ *and* $l=(1-\alpha)\ln n+o(\ln n)$ *as* $n\to\infty$, *where* $0<\alpha<1$, *then for any fixed* t,

$$\varphi_{(r)}^{l}\left(\frac{t}{n}\right)=\frac{1}{1-it}+o(1).$$

PROOF. Using the representation (2) and the estimates (3) and (4), we obtain

$$\varphi_n\left(\frac{t}{n}\right)=1-\frac{\ln(1-it)}{\ln n}+O\left(\frac{1+|t|}{n\ln n}\right). \tag{5}$$

Just as in Lemma 5, we find here that

$$\sum_{k=1}^{r}p_k e^{itk/n}=\sum_{k=1}^{r}\frac{\left(1-\frac{1}{n}\right)^k e^{itk/n}}{k\ln n}=$$

$$=\frac{1}{\ln n}\sum_{k=1}^{r}\frac{1}{k}+O\left(\frac{r(1+|t|)}{n\ln n}\right)=\alpha+O\left(\frac{1+|t|}{\sqrt{\ln n}}\right). \tag{6}$$

Using equation (5), the estimate given in Lemma 5, and equation (6), we obtain for any fixed t

$$\varphi_{(r)}^{l}\left(\frac{t}{n}\right)=\left(1-\frac{\ln(1-it)}{\ln n}-\frac{1}{\ln n}\sum_{k=1}^{r}\frac{1}{k}+O\left(\frac{r(1+|t|)}{n\ln n}\right)\right)^{l}\times$$

$$\times\left(1-\frac{1}{\ln n}\sum_{k=1}^{r}\frac{1}{k}+O\left(\frac{r}{n\ln n}\right)\right)^{-l}=$$

$$=\left(1-\frac{\ln(1-it)}{(1-\alpha)\ln n}+o\left(\frac{1}{\ln n}\right)\right)^{(1-\alpha)\ln n+o(\ln n)}=\frac{1}{1-it}+o(1).$$

As follows from Lemma 6, the distribution of the normalized sums $\zeta_l^{(A_r)}/n$ as $n\to\infty$, where $l=(1-\alpha)\ln n+o(\ln n)$ and $r=\exp\{\alpha\ln n+o(\ln n)\}$ converges to an exponential distribution.

Following the proof of Theorem 1 and replacing the integration domains $A\leqslant|t|\leqslant\varepsilon n$ and $\varepsilon n<|t|\leqslant\pi n$ with $A\leqslant|t|\leqslant n/r$ and $n/r<|t|\leqslant\pi n$ respectively in the definition of the integrals I_2 and I_3, we can easily prove the local theorem for $\zeta_l^{(A_r)}/n$:

Theorem 3. *If* $l=(1-\alpha)\ln n+o(\ln n)$ *and* $r=\exp\{\alpha\ln n+o(\ln n)\}$, *where* $0<\alpha<1$, *as* $n\to\infty$ *and if* $z=k/n$, *where* k *are natural numbers, then*

$$n\mathbf{P}\left\{\frac{1}{n}\zeta_l^{(A_r)}=z\right\}=e^{-z}+o(1)$$

uniformly with respect to $z\geqslant z_0>0$.

For $\zeta_l^{(A_r)}/n$, we need an estimate that, though cruder, holds for all natural numbers l.

Lemma 7. *If* $n>l$ *and* $r=\exp\{\alpha\ln n+o(\ln n)\}$, *where* α *is a fixed number in* (0, 1), *there will be a constant* c *such that, starting from some* n,

$$\mathbf{P}\left\{\zeta_l^{(A_r)}=n\right\}\leqslant\frac{cl^2}{n\ln n}.$$

PROOF. Since

$$\left\{\zeta_l^{(A_r)}=n\right\}=\bigcup_{i=1}^{l}\bigcup_{k\geqslant[n/l]}\left\{\xi_i^{(A_r)}=k,\ \zeta_l^{(A_r)}=n\right\},$$

we have

$$\mathbf{P}\left\{\zeta_l^{(A_r)}=n\right\}\leqslant l\sum_{k\geqslant[n/l]}\mathbf{P}\left\{\xi_1^{(A_r)}=k,\ \zeta_l^{(A_r)}=n\right\}=$$

$$=l\sum_{k\geqslant[n/l]}\mathbf{P}\left\{\xi_1^{(A_r)}=k\right\}\mathbf{P}\left\{\zeta_{l-1}^{(A_r)}=n-k\right\}\leqslant$$

$$\leqslant\frac{lp_{[n/l]}}{1-\mathbf{P}\{\xi_1\leqslant r\}}\sum_{k\geqslant[n/l]}\mathbf{P}\left\{\zeta_{l-1}^{(A_r)}=n-k\right\}\leqslant\frac{lp_{[n/l]}}{1-\mathbf{P}\{\xi_1\leqslant r\}},$$

from which the assertion of the lemma easily follows.

Let us take now the sums $\zeta_l^{(A_r)}+\zeta_{N-l}^{(\overline{A}_r)}$ for $r=\gamma n$, where $0<\gamma<1$. We introduce the function

$$E(z)=\int_z^\infty\frac{1}{u}e^{-u}\,du,$$

where $\operatorname{Re} z>0$ and the integration is over the path consisting of the segment $[z,\ \operatorname{Re} z]$ and the ray $[\operatorname{Re} z,\ +\infty]$. We estimate first

$$S_{\gamma n}(t)=\sum_{k>\gamma n}\frac{1}{k}\left(1-\frac{1}{n}\right)^k e^{\frac{itk}{n}}.$$

Lemma 8. *For any fixed* γ *such that* $0<\gamma<1$, *we have*

$$S_{\gamma n}(t)=E(\gamma(1-it))+o(1).$$

as $n\to\infty$.

PROOF. Let us note first that we can write

$$S_{\gamma n}(t) = \sum_{k>\gamma n} \frac{1}{k}\left(\left(1-\frac{1}{n}\right)e^{it/n}\right)^k =$$

$$= \sum_{k>\gamma n} \int_0^{(1-n^{-1})e^{it/n}} x^{k-1}\,dx = \int_0^{(1-n^{-1})e^{it/n}} \frac{x^{[\gamma n]}}{1-x}\,dx.$$

Making the substitution, $y=n(1-x)$ we get

$$S_{\gamma n}(t) = \int_{n-(n-1)e^{it/n}}^{n} \frac{1}{y}\left(1-\frac{y}{n}\right)^{[\gamma n]} dy.$$

The integrand is majorized by an integrable function. Hence, by taking the limit under the integral sign, we find that

$$S_{\gamma n}(t) = \int_{1-it}^{\infty} \frac{e^{-\gamma y}}{y}\,dy + o(1) = \int_{\gamma(1-it)}^{\infty} \frac{1}{u}e^{-u}\,du + o(1).$$

In particular, we obtain the estimate

$$\mathbf{P}\{\xi_1 > \gamma n\} = \frac{1}{\ln n} S_{\gamma n}(0) = \frac{E(\gamma)+o(1)}{\ln n} \tag{7}$$

for the tail of the distribution of the logarithmic series.

Theorem 4. *If* $N=\ln n+o(\ln n)$ *as* $n\to\infty$, *where* l *is a fixed natural number, and if* $z=k/n$, *where* k *is a natural number, then*

$$n\mathbf{P}\left\{\frac{1}{n}\left(\zeta_l^{(A\ \gamma n)}+\zeta_{N-l}^{(\bar{A}\gamma n)}\right)=z\right\}=$$

$$=\frac{e^{-z}}{E^l(\gamma)\,e^{-E(\gamma)}}\sum_{s=0}^{\infty}\frac{(-1)^s}{s!}\int_{X(s+l,z,\gamma)}\frac{dx_1\ldots dx_{s+l}}{x_1\ldots x_{s+l}}+o(1) \tag{8}$$

uniformly with respect to $z \geqslant z_0 > 0$, *where* $X(s, z, \gamma) = \{x_i \geqslant \gamma,\ i=1, \ldots, s,\ x_1 + \ldots + x_s \leqslant z\}$ (*for* $s+l=0$, *the integral should be assumed to be equal to one*).

PROOF. Let us look at the left-hand member of (8). The characteristic function of the term $\zeta_{N-l}^{(\bar{A}_{\gamma n})}/n$ by itself is equal to

$$f_1(t) = \frac{\varphi_n\left(\frac{t}{n}\right) - \frac{1}{\ln n} S_{\gamma n}(t)}{1 - \frac{1}{\ln n} S_{\gamma n}(0)}.$$

Using the estimates given in Lemmas 1 and 8, we easily find that

$$f_1(t) = \left(1 - \frac{\ln(1-it) + E(\gamma(1-it)) + o(1)}{\ln n}\right) \times \\ \times \left(1 - \frac{E(\gamma) + o(1)}{\ln n}\right)^{-1}$$

as $n \to \infty$ from which it follows that, for any fixed l, if $N = \ln n + o(\ln n)$ as $n \to \infty$, then

$$f_1^{N-l}(t) = \frac{e^{-E(\gamma(1-it))}}{(1-it)\, e^{-E(\gamma)}} + o(1).$$

The characteristic function of the term $\zeta_l^{(A_{\gamma n})}/n$ in (8) by itself is equal to $f_2(t) = \dfrac{S_{\gamma n}(t)}{S_{\gamma n}(0)}$ and

$$f_2(t) = \frac{E(\gamma(1-it))}{E(\gamma)} + o(1)$$

as $n \to \infty$.

Hence the characteristic function $\varphi_{l,\gamma}(t)$ of the sum $\left(\zeta_l^{(\bar{A}_{\gamma n})} + \zeta_{N-l}^{(\bar{A}_{\gamma n})}\right)/n$ satisfies the relation

$$\varphi_{l,\gamma}(t) = \frac{E^l(\gamma(1-it))\, e^{-E(\gamma(1-it))}}{E^l(\gamma)\, e^{-E(\gamma)}(1-it)} + o(1)$$

for any fixed l as $n \to \infty$.

Since $E(\gamma(1-it))$ is the Fourier transform of the function

$$f(x)=\begin{cases}\dfrac{1}{x}e^{-x} & \text{for } x>\gamma,\\[2ex] 0 & \text{for } x<\gamma,\end{cases}$$

if we expand $e^{-E(\gamma(1-it))}$ as a series of powers of $E(\gamma(1-it))$, we find that $\varphi_{l,\gamma}(t)$ converges to the characteristic function of a distribution with density

$$f_{l,\gamma}(z)=\frac{e^{-z}}{E^{l}(\gamma)\,e^{-E(\gamma)}}\sum_{s=0}^{\infty}\frac{(-1)^{s}}{s!}\int\limits_{X(s+l,z,\gamma)}\frac{dx_1\ldots dx_{s+l}}{x_1\ldots x_{s+l}},$$

where the integral should be taken equal to one for $s+l=0$. The theorem is then proved in a similar way as Theorem 1, because the inversion formula holds for $\varphi_{l,\gamma}(t)$ and the assertions of Lemmas 3, 4, and 5 hold for the characteristic function of the term $\zeta_{N-l}^{(\bar{A}_{\gamma n})}/n$ in (8) by itself.

§4. Cycles of a Random Permutation

We shall apply the results of the previous section to the study of random variables related to the cyclic structure of a permutation.

Theorem 1.

$$\mathbf{P}\{\varkappa_n=N\}=\frac{1}{\sqrt{2\pi\ln n}}\exp\left\{-\frac{(N-\ln n)^2}{2\ln n}\right\}(1+o(1))$$

uniformly with respect to $(N-\ln n)/(\ln n)^{7/12}$ *in any finite interval as* $n\to\infty$.

PROOF. By Lemma 2.1, for $\theta=1-\frac{1}{n}$ we have

$$\mathbf{P}\{\varkappa_n=N\}=\frac{(\ln n)^N}{N!\left(1-\frac{1}{n}\right)^n}\mathbf{P}\{\zeta_N=n\}.$$

One can easily see that the assertion of Theorem 3.1 holds uniformly with respect to values of N for which $(N-\ln n)/(\ln n)^{7/12}$ lies in any finite interval. Hence, applying Theorem 3.1, we see that uniformly with respect to $(N-\ln n)/(\ln n)^{7/12}$ in any finite interval

$$\mathbf{P}\{\varkappa_n = N\} = \frac{(\ln n)^N}{N!n}(1+o(1)),$$

which implies the statement of the theorem.

We consider next the joint distribution of the random variables $\alpha_{r_1}, \alpha_{r_2}, \ldots, \alpha_{r_s}$, where $r_1, r_2, \ldots, r_s$ are natural numbers.

Theorem 2. *For any natural numbers* $k_1, \ldots, k_s$,

$$\mathbf{P}\{\alpha_{r_1} = k_1, \ldots, \alpha_{r_s} = k_s\} = \\ = \frac{r_1^{-k_1} \ldots r_s^{-k_s}}{k_1! \ldots k_s!} \exp\left\{-\frac{1}{r_1} - \ldots - \frac{1}{r_s}\right\} + o(1)$$

as $n \to \infty$.

PROOF. By Theorem 2.3,

$$\mathbf{P}\{\alpha_{r_1} = k_1, \ldots, \alpha_{r_s} = k_s\} = \\ = \sum_{N=1}^{n} \mathbf{P}\{\nu_n = N\} \mathbf{P}\{\mu_{r_1}(n, N) = k_1, \ldots, \mu_{r_s}(n, N) = k_s\}. \quad (1)$$

By Theorems 3.1 and 3.2, if $N = \ln n + o(\ln n)$ as $n \to \infty$, then

$$\frac{\mathbf{P}\{\zeta^{(r_1,\ldots,r_s)}_{N-k_1-\ldots-k_s} = n - k_1 r_1 - \ldots - k_s r_s\}}{\mathbf{P}\{\zeta_N = n\}} = 1 + o(1). \quad (2)$$

Hence, by Lemma 1.1, if $N = \ln n + o(\ln n)$ as $n \to \infty$, we have

$$\mathbf{P}\{\mu_{r_1}(n, N) = k_1, \ldots, \mu_{r_s}(n, N) = k_s\} = \\ = \frac{1}{k_1! \ldots k_s! r_1^{k_1} \ldots r_s^{k_s}} \exp\left\{-\frac{1}{r_1} - \ldots - \frac{1}{r_s}\right\} + o(1).$$

Since these equalities are satisfied uniformly with respect to $(N-\ln n)/(\ln n)^{7/12}$ in any finite interval, by applying Theorem 3.1 we obtain from (1) the assertion of the theorem.

The following statement on the distribution of the lengths of mth minimum cycles S_m is a simple corollary of this theorem and the identity

$$\mathbf{P}\{S_m \leqslant r\} = \mathbf{P}\{\alpha_1 + \ldots + \alpha_r \geqslant m\},$$

which holds for all natural numbers m and r.

Corollary 1. *For any natural number* r,

$$\mathbf{P}\{S_m \leqslant r\} = \sum_{k=m}^{\infty} \frac{\lambda_r^k e^{-\lambda_r}}{k!} + o(1),$$

where $\lambda_r = 1 + \frac{1}{2} + \ldots + \frac{1}{r}$.

From Theorem 2 and the identity

$$\mathbf{P}\{S_1 = r,\ \alpha_r = s\} = \mathbf{P}\{\alpha_1 + \ldots + \alpha_{r-1} = 0,\ \alpha_r = s\},$$

which holds for natural numbers r and s, the following statement on the number of cycles of minimum length follows.

Corollary 2. *For all natural numbers* r *and* s,

$$\mathbf{P}\{S_1 = r, \alpha_r = s\} = \frac{1}{s!r^s} e^{-\lambda_r} + o(1),$$

where $\lambda_r = 1 + \frac{1}{2} + \ldots + \frac{1}{r}$.

The statement which follows holds true for the random variables $S_{\varkappa_n - m+1}$ equal to the length of the mth maximum cycle.

Theorem 3. *As* $n \to \infty$,

$$\mathbf{P}\left\{\frac{1}{n} S_{\varkappa_n - m+1} \leqslant \gamma\right\} =$$

$$= \begin{cases} 1 - \sum\limits_{m \leqslant s < 1/\gamma} \dfrac{(-1)^{s-m}}{s\,(m-1)!\,(s-m)!} \underset{\substack{x_1+\ldots+x_s \leqslant 1 \\ \min x_i \geqslant \gamma}}{\int \cdots \int} \dfrac{dx_1 \ldots dx_s}{x_1 \ldots x_s} + \\ \qquad + o(1), \quad 0 < \gamma < \dfrac{1}{m}, \\ 1, \qquad \gamma \geqslant \dfrac{1}{m}. \end{cases}$$

PROOF. First we consider the distribution of $\eta_{(N-m+1)}$. By Lemma 1.2, for $0<\gamma<1$,

$$\mathbf{P}\left\{\frac{1}{n} \eta_{(N-m+1)} \leqslant \gamma\right\} = \sum_{l=0}^{m-1} C_N^l (\mathbf{P}\{\xi_1 > \gamma n\})^l \times$$
$$\times (\mathbf{P}\{\xi_1 \leqslant \gamma n\})^{N-l} \frac{\mathbf{P}\left\{\zeta_l^{(A_{\gamma n})} + \zeta_{N-l}^{(\overline{A}_{\gamma n})} = n\right\}}{\mathbf{P}\{\zeta_N = n\}}. \quad (3)$$

If $N = \ln n + o(\ln n)$ as $n \to \infty$, we have by Theorem 3.1 $n\mathbf{P}\{\zeta_N = n\} = e^{-1} + o(1)$ and by Theorem 3.4

$$n\mathbf{P}\left\{\zeta_l^{(A_{\gamma n})} + \zeta_{N-l}^{(\overline{A}_{\gamma n})} = n\right\} =$$
$$= \frac{e^{-1}}{E^l(\gamma)\, e^{-E(\gamma)}} \sum_{s=0}^{\infty} \frac{(-1)^s}{s!} \int\limits_{X(s+l,1,\gamma)} \frac{dx_1 \ldots dx_{s+l}}{x_1 \ldots x_{s+l}} + o(1),$$

where the domain $X(s+l,\ 1,\ \gamma)$ is the same as in Theorem 3.4 and the integral is defined to be equal to one if $s+l=0$; moreover, by virtue of (3.7),

$$\mathbf{P}\{\xi_1 > \gamma n\} = \frac{E(\gamma) + o(1)}{\ln n}$$

and hence $(1-\mathbf{P}\{\xi_1 > \gamma n\})^{N-l} = e^{-E(\gamma)} + o(1)$. Substituting these expressions into (3), we find

$$\mathbf{P}\left\{\frac{1}{n}\eta_{(N-m+1)}\leqslant\gamma\right\}=$$
$$=\sum_{l=0}^{m-1}\frac{1}{l!}\sum_{s=0}^{\infty}\frac{(-1)^s}{s!}\int\limits_{X(s+l,1,\gamma)}\frac{dx_1\ldots dx_{s+l}}{x_1\ldots x_{s+l}}+o(1),$$

From this, using the identity

$$\sum_{l=0}^{m-1}(-1)^l C_n^l=\begin{cases}(-1)^{m-1}C_{n-1}^{m-1}, & n\geqslant m,\\ 0, & n<m,\end{cases}$$

we get

$$\mathbf{P}\left\{\frac{1}{n}\eta_{(N-m+1)}\leqslant\gamma\right\}=$$
$$=1-\sum_{m\leqslant s<1/\gamma}\frac{(-1)^{s-m}}{s(m-1)!\,(s-m)!}\int\limits_{X(s,1,\gamma)}\frac{dx_1\ldots dx_s}{x_1\ldots x_s}+o(1).$$

This relation is satisfied uniformly with respect to $(N-\ln n)/(\ln n)^{7/12}$ in any finite interval. Hence, averaging with respect to the distribution of ν_n and using Theorem 1, we arrive at the assertion of the theorem.

Let us consider now the distribution of the mean terms of the order sequence of the lengths of cycles of a random permutation. Since the average number of the cycles is of the order of $\ln n$ as $n\to\infty$ the mean terms of S_m are equal to m of the order of $\alpha\ln n$, where $0<\alpha<1$. For the lengths of these cycles we have the following theorem:

Theorem 4. *If* $m=\alpha\ln n+o(\sqrt{\ln n})$, *where* $0<\alpha<1$, *as* $n\to\infty$, *then*

$$\mathbf{P}\left\{\frac{\ln S_m-m}{\sqrt{m}}\leqslant t\right\}=\frac{1}{\sqrt{2\pi}}\int\limits_{-\infty}^{t}e^{-u^2/2}du+o(1).$$

PROOF. Let us consider the probability

$$\mathbf{P}\left\{\frac{\ln S_m - \alpha \ln n}{\sqrt{\alpha \ln n}} \leqslant t\right\} = \mathbf{P}\{S_m \leqslant r\},$$

where $r = \exp\{\alpha \ln n + t\sqrt{\alpha \ln n}\}$. For estimating $\mathbf{P}\{S_m \leqslant r\}$, we first find the asymptotic behavior of the probability $\mathbf{P}\{\eta_{(m)} \leqslant r\}$ and then average the result with respect to the distribution of ν_n. Let us use (1.5) of Lemma 1.2. Let us set $N = \ln n + s\sqrt{\ln n}$, where $|s| < (\ln n)^{1/12}$ and $l = \alpha \ln n + u\sqrt{\alpha(1-\alpha)\ln n}$, and let us divide the summation domain into two parts:

$$L_1 = \{0 \leqslant l < \alpha \ln n - 2(\ln n)^{7/12}\},$$
$$L_2 = \{\alpha \ln n - 2(\ln n)^{7/12} \leqslant l \leqslant m-1\}.$$

Since $\zeta_l^{(\bar{A}_r)} \leqslant lr$ and $|s| < (\ln n)^{1/12}$, by Lemma 3.7 we infer that

$$\mathbf{P}\left\{\zeta_l^{(\bar{A}_r)} + \zeta_{N-l}^{(A_r)} = n\right\} \leqslant \frac{c \ln n}{n}$$

and by Theorem 3.1,

$$n\mathbf{P}\{\zeta_N = n\} = e^{-1} + o(1).$$

Hence in equation (1.5) of Lemma 1.2,

$$\frac{\mathbf{P}\left\{\zeta_l^{(\bar{A}_r)} + \zeta_{N-l}^{(A_r)} = n\right\}}{\mathbf{P}\{\zeta_N = n\}} = O(\ln n).$$

Making use of the normal approximation for the tail of a binomial distribution and taking into account the fact that, by Lemma 3.5,

$$\mathbf{P}\{\xi_1 \leqslant r\} = \alpha + \frac{t\sqrt{\alpha}}{\sqrt{\ln n}} + O\left(\frac{1}{\ln n}\right),$$

for sufficiently large n we obtain the estimate

$$\sum_{l\in L_1} C_N^l\,(\mathbf{P}\{\xi_1\leqslant r\})^l\,(\mathbf{P}\{\xi_1>r\})^{N-l}\,\frac{\mathbf{P}\left\{\zeta_l^{(\overline{A}_r)}+\zeta_{N-l}^{(A_r)}=n\right\}}{\mathbf{P}\{\zeta_N=n\}}\leqslant$$

$$\leqslant\frac{2c\ln n}{\sqrt{2\pi}}\int\limits_{-\infty}^{-2a(\ln n)^{1/12}-ab}e^{-u^2/2}du\to 0,$$

where $a=(\alpha(1-\alpha))^{-1/2}$ and $b=\alpha s+t\sqrt{\alpha}.$

In the domain L_2, we infer from Theorems 3.1 and 3.3 that

$$\frac{\mathbf{P}\left\{\zeta_l^{(\overline{A}_r)}+\zeta_{N-l}^{(A_r)}=n\right\}}{\mathbf{P}\{\zeta_N=n\}}=1+o(1)$$

uniformly with respect to $|s|<(\ln n)^{1/12}$ and u such that $l\in L_2$. Hence, using the normal approximation for a binomial distribution, we have

$$1-\sum_{l\in L_2}C_N^l(\mathbf{P}\{\xi_1\leqslant r\})^l(\mathbf{P}\{\xi_1>r\})^{N-l}\,\frac{\mathbf{P}\left\{\zeta_l^{(\overline{A}_r)}+\zeta_{N-l}^{(A_r)}=n\right\}}{\mathbf{P}\{\zeta_N=n\}}=$$

$$=1-\frac{1}{\sqrt{2\pi}}\int\limits_{-\infty}^{-ab}e^{-u^2/2}du+o(1)=\frac{1}{\sqrt{2\pi}}\int\limits_{-\infty}^{ab}e^{-u^2/2}\,du+o(1).$$

Averaging over the distribution of ν_n, we obtain

$$\mathbf{P}\{S_m\leqslant r\}=\sum_{N=1}^{n}\mathbf{P}\{\eta_{(m)}\leqslant r\}\,\mathbf{P}\{\nu_n=N\}=$$

$$=\frac{1}{2\pi}\int\limits_{-\infty}^{\infty}e^{-s^2/2}\int\limits_{-\infty}^{ab}e^{-u^2/2}du\,ds+o(1)=$$

$$=\frac{1}{\sqrt{2\pi}}\int\limits_{-\infty}^{t}e^{-u^2/2}du+o(1),$$

thus proving the theorem.

§5. Further Results. References

A probabilistic approach to the investigation of the properties of the class S_n of permutations of increasing degree n was first employed by Goncharov [15, 16]. Using the generating function (4.2), he obtained the basic results on the limit behavior of random variables associated with the cyclic structure of a random permutation. In [16], he proved an integral theorem on the convergence of the distribution of the number of cycles $\varkappa_n$ of a random permutation to a normal distribution with parameters $(\ln n, \ln n)$; it is also shown there that $\mathbf{M}\varkappa_n = \ln n + c + o(1)$ (where c is Euler's constant); a limit theorem on convergence to the Poisson distribution with parameter $1/r$ is proved for the number α_r of cycles of length r. In this paper, the limit distribution of the maximum term $S_{\varkappa_n}$ of the order sequence of lengths of cycles is found.

Even though the cyclic structure of random permutations had been investigated rather completely by Goncharov, the subject still attracted the attention of many other mathematicians. An elegant approach to the study of the class of random permutations was taken in Shepp and Lloyd [96]. Let $\xi_1, \ldots, \xi_n$ be independent Poisson-distributed random variables with $\mathbf{M}\xi_r = z^r/r!$ (where z is a parameter). Then, as shown in [96],

$$\mathbf{P}\{\alpha_r = k_r,\ r = 1, \ldots, n\} = \\ = \mathbf{P}\{\xi_r = k_r,\ r = 1, \ldots, n \mid \xi_1 + 2\xi_2 + \ldots + n\xi_n = n\}.$$

Exploiting the relation between the random variables $\alpha_1, \ldots, \alpha_n$ describing the cyclic structure of a random permutation and the independent random variables $\xi_1, \ldots, \xi_n$, the authors investigate in [96] the limit behavior of the extreme terms of the order sequence $S_1, \ldots, S_{\varkappa_n}$. Asymptotic expressions are obtained for distributions of these random variables as well as for their moments of any order. In particular, the value 0.6243... is obtained for $\lim\limits_{n\to\infty} \frac{1}{n} \mathbf{M}S_{\varkappa_n}$. The ideas presented in [96] were developed and generalized by

Balakrishnan, Sankaranarayanan, and Suyambulingom in [57].

Another approach to the study of the limit behavior of the maximum term of an order sequence was taken by Stepanov in [49]; here one finds a summary of results on random mappings of finite sets, as well as on the limit behavior of characteristics of random permutations.

A new approach to the investigation of asymptotic properties of a set of one-to-one mappings with a uniform distribution is suggested by Vershik and Schmidt [8]. This approach differs from all previous approaches mainly in the fact that it makes it possible to indicate a space of elementary events on which probability measures induced by a uniform distribution on S_n converge weakly to a limit distribution. Therefore, one can compute limit distributions of various characteristics of a random permutation by using this limit distribution. We shall describe here one of the constructions given in that paper. Let us denote by $x_i(T)$ the relative length of the ith (in decreasing order) cycle of a permutation T. The mapping $\{x_1(T), \ldots, x_n(T), 0, 0, \ldots\}$ of a set S_n with a uniform distribution onto the set Γ of all sequences $\{x_i\}$ with nonnegative nonincreasing elements the sum of which is equal to one generates a probability distribution ν_n on Γ. The main result consists in the fact that on Γ there exists a probability distribution ν, which is a weak limit for ν_n. This fact enables us to obtain many familiar results on the limit distributions of the characteristics of random permutations.

The approach used in Chapter 8 to the study of the asymptotic properties of random permutations is discussed by Kolchin [29, 30]. In [31], with the help of this approach, a local theorem on convergence to a normal distribution is obtained for the distribution of $\varkappa_n$, and the limit distributions of the middle terms of the order sequence of lengths of the cycles of a random permutation are investigated.

The study of the group properties of a random permutation T as an element of a symmetric group S_n is of great interest. The investigations made in this area are not as definitive as the results related to the behavior of elements of the set $\{1, 2, \ldots, n\}$ under a random mapping. The investigations of the order $O_n(T)$ of a random permutation of degree n are most typical. (The order of a permutation T is equal to the least common multiple of the lengths of the cycles of that permutation.) Erdös and Turan examine in [74] the limit behavior of the number $k(T)$ of cycles of various lengths in a random permutation. They show in [75] that the random variable

$$\frac{\ln O_n(T) - \frac{1}{2}\ln^2 n}{\sqrt{\frac{1}{3}\ln^3 n}}$$

is asymptotically normal with parameters (0, 1). Another proof of asymptotic normality of $\ln O_n(T)$ can be found in Best [64]. Local theorems on the number of permutations $T \in S_n$ of simple order p are presented in Chowla, Herstein, and Moore [66] and also in Moser and Wyman [87]. It will be noted that the relation

$$\lim_{n\to\infty} \frac{\ln G(n)}{\sqrt{n \ln n}} = 1.$$

holds true for the maximum order $G(n) = \max_{T \in S_n} O_n(T)$.

The probabilistic approach to the problem of generating a group by the elements of that group is of some interest. We denote by P_n the probability that the group generated by two random permutations of S_n will coincide with S_n. Dixon [71] proved that $P_n \to 0.75$ as $n \to \infty$.

Great difficulties arise in considering a class of permutations S_n with a distribution different from a uniform distribution. We mention in this respect Sachkov [42] and Tarakanov and Chistyakov [50], where linear combinations $\alpha_1 + 2\alpha_2 + \ldots + r\alpha_r$ (for fixed r) for some subclasses of S_n with a uniform distribution are studied.

BIBLIOGRAPHY

1. Belyayev, P. F., On probability of nonoccurrence of a given number of outcomes, *Teoriya veroyatn. i yeye primen.,* **9**, No. 9, 541–547 (1964).
2. Belyayev, P. F., On probability of nonoccurrence of a given number of s-chains in compound Markov chains, *Teoriya veroyatn. i yeye primenen.,* **10**, No. 3, 547–551 (1965).
3. Belyayev, P. F., On the joint distribution of frequences of long s-chains in a multinomial scheme with equiprobable outcomes, *Teoriya veroyatn. i yeye primenen.,* **14**, No. 3, 540–546 (1969).
4. Bernstein, S. N., *Teoriya veroyatnostey* (Probability theory), 4th ed., Moscow and Leningrad, 1946.
5. Bolotnikov, Yu. V., Convergence of variables $\mu_r(n, N)$ to a Gaussian or a Poisson process in the classical shot problem, *Teoriya veroyatn. i yeye primenen.,* **13**, No. 1, 39–50 (1968).
6. Bolotnikov, Yu. V., Convergence of a number of empty cells to a Gaussian process in the classical problem of allocation of particles to cells, *Matem. zametki,* **4**, No. 1, 97–103 (1968).
7. Bolotnikov, Yu. V., Limit processes in an equiprobable scheme of allocation of particles to cells, *Teoriya veroyatn. i yeye primenen.,* **13**, No. 3, 534–542 (1968).

8. Vershik, A. M., and A. A. Schmidt, Symmetric groups of higher degree, *Dokl. Akademii Nauk SSSR,* **206**, No. 2, 269–272 (1972).
9. Viktorova, I. I., On the asymptotic behavior of the maximum in an equiprobable polynomial scheme, *Matem. zametki,* **5**, No. 3, 305–316 (1969).
10. Viktorova, I. I., The empty cell test and its generalizations, Candidate's dissertation, 1971.
11. Viktorova, I. I., and V. A. Sevast'yanov, On the limit behavior of the maximum in a polynomial scheme, *Matem zametki,* **1**, No. 3, 331–338 (1967).
12. Viktorova, I. I., and V. P. Chistyakov, Asymptotic normality in the problem of shot with arbitrary probabilities of falling into the cells, *Teoriya veroyatn. i yeye primenen.*, **10**, No. 1, 162–167 (1965).
13. Viktorova, I. I., and V. P. Chistyakov, Some generalizations of the empty cell test, *Teoriya veroyatn. i yeye primenen.,* **11**, No. 2, 306–313 (1966).
14. Gikhman, I. I., and A. V. Skorokhod, *Introduction to the Theory of Random Processes*, Philadelphia, Saunders, 1969 (translation of Vvedeniye v. teoriyu sluchaynykh protsessov).
15. Goncharov, V. L., On the distribution of cycles in permutations, *Dokl. Akademii Nauk SSSR,* **35**, No. 9, 299–310 (1942).
16. Goncharov, V. L., Some facts from combinatorics, *Izvestia Akademii Nauk SSSR, Ser. Matem.*, **8**, No. 1, 3–48 (1944).
17. Yermakov, S. M., *Metod Monte-Karlo i smezhnyye voprosy*, (Monte-Carlo simulations and related problems), Izd. Nauka, Moscow, 1971.
18. Zubkov, A. M., and V. G. Mikhaylov, Limit distributions of random variables associated with long recurrences in a sequence of independent trials, *Teoriya veroyatn. i yeye primenen.,* **19**, No. 1, 173–181 (1974).
19. Ivchenko, G. I., On limit distributions of order statistics of a polynomial scheme, *Teoriya veroyatn. i yeye primenen.,* **16**, No. 1, 94–107 (1971).
20. Ivchenko, G. I., Limit theorems in a problem of allocation, *Teoriya veroyatn. i yeye primenen.,* **16**, No. 2, 292–305 (1971).
21. Ivchenko, G. I., An order sequence for a scheme of summation of independent variables, *Teoriya veroyatn. i yeye primenen.,* **18**, No. 3, 557–570 (1973).

22. Ivchenko, G. I., On extremal values of samples, *Trudy Moscov. Instituta electron. mashinostr.*, **32**, 12–31 (1973).
23. Ivchenko, G. I., Expectation time and an order sequence of frequencies in a polynomial scheme, *Trudy Moscov. Instituta electron. mashinostr.*, **32**, 39–64 (1973).
24. Ivchenko, G. I., and Yu. I. Medvedev, Some multidimensional theorems in the classical problem of allocation, *Teoriya veroyatn. i primenen.*, **10**, No. 1, 156–162 (1965).
25. Ivchenko, G. I., and Yu. I. Medvedev, Asymptotic behavior of the number of particle sets in the classical problem of allocation, *Teoriya veroyatn. i yeye primenen.*, **11**, No. 4, 701–708 (1966).
26. Ivchenko, G. I., Yu. I. Medvedev, and B. A. Sevast'yanov, Allocation of a random number of particles to cells, *Matem. zametki,* **1**, No. 5, 549–554 (1967).
27. Kolchin, V. F., The rate of convergence to limit distributions in the classical problem of shots, *Teoriya veroyatn. i yeye primenen.*, **11**, No. 1, 144–156 (1966).
28. Kolchin, V. F., A case of uniform local limit theorems with a variable lattice in the classical problem of shot, *Teoriya veroyatn. i yeye primenen.*, **12**, No. 1, 62–72 (1967).
29. Kolchin, V. F., One class of limit theorems for conditional distributions, *Liet. mat. rinkinys,* **8**, No. 1, 53–63 (1968).
30. Kolchin, V. F., On the limit behavior of the extremal terms of an order sequence in a polynomial scheme, *Teoriya veroyatn. i yeye primenen.*, **14**, No. 3, 476–487 (1969).
31. Kolchin, V. F., A problem of allocation of particles to cells and cycles of random permutations, *Teoriya veroyatn. i yeye primenen.*, **16**, No. 1, 67–82 (1971).
32. Kolchin, V. F., On the distribution of one statistic in a polynomial scheme, *Trudy Moscov. Instituta electron. mashinostr.*, **32**, 73–91 (1973).
33. Kolchin, V. F., and V. P. Chistyakov, New limit distributions in a problem of allocation, *Trudy Moscov. Instituta electron. mashinostr.*, **32**, 65–72 (1973).
34. Kolchin, V. F. and V. P. Chistyakov, Combinatorial problems in probability theory, *VINITI, Itogi nauki i tekhniki, seria: Teoriya veroyat., Matemat. statistika, Teoret. kibernetika,* **11**, 5–45 (1974).
35. Kolchin, V. F., and V. P. Chistyakov, Limit distributions of the number of failed *s*-chains in a polynomial scheme, *Teoriya veroyatn. i yeye primenen.*, **19**, No. 4, 855–864 (1974).

36. Medvedev, Yu. I., Some theorems on the asymptotic distribution of the χ^2 statistic, *Dokl. Akademii Nauk SSSR,* **192**, No. 5, 987–989 (1970).
37. Mikhaylov, V. G., Limit distributions of random variables associated with many long recurrences in a sequence of independent trials, *Teoriya veroyatn. i yeye primenen.,* **19**, No. 1, 182–187 (1974).
38. Mikhaylov, V. G., Convergence to a Poisson process in a scheme of increasing sums of dependent random variables, *Teoriya veroyatn. i yeye primenen.,* **19**, No. 2, 422–426 (1974).
39. Mikhaylov, V. G., Convergence to a process with independent increments in a scheme of increasing sums of dependent random variables, *Matem. sbornik,* **94**, (136), No. 2 (6), 283–299 (1974).
40. Popova, T. Yu., Limit theorems in one model of allocation of two types of particles, *Teoriya veroyatn. i yeye primenen.,* **13**, No. 3, 542–548 (1968).
41. Prokhorov, Yu. V., Asymptotic behavior of a binomial distribution, *Uspekhi matem. nauk,* **8**, No. 3, 135–142 (1953).
42. Sachkov, V. N., On extreme points of the space of symmetric stochastic matrices, *Matem. sbornik,* **96** (138), No. 3, 447–457 (1975).
43. Sevast'yanov, B. A., Limit theorems in a scheme of allocation of particles to cells, *Teoriya veroyatn. i yeye primenen.,* **11**, No. 4, 696–700 (1966).
44. Sevast'yanov, B. A., Convergence of the number of empty boxes in the classical problem of shot to a Gaussian or a Poisson process, *Teoriya veroyatn. i yeye primenen.,* **12**, No. 1, 144–154 (1967).
45. Sevast'yanov, B. A., The empty cell test and its generalizations, *Trudy Instituta prikl. matem. Tbilissk. un-ta,* **No. 2**, **733**–738 (1969).
46. Sevast'yanov, B. A., A limiting Poisson law in a scheme of sums of dependent random variables, *Teoriya veroyatn. i yeye primenen.,* **17**, No. 4, 733–738 (1972).
47. Sevast'yanov, B. A., and V. P. Chistyakov, Asymptotic normality in the classical problem of shot, *Teoriya veroyatn. i yeye primenen.,* **9**, No. 2, 223–237 (1964); letter to the editor, **9**, No. 3, 568 (1964).
48. Sobol', I. M., The Monte Carlo Method, University of Chicago Press, 1975 (translation of *Chislennyye metody Monte-Karlo*).

49. Stepanov, V. Ye., Limit distributions of some characteristics of random mappings, *Teoriya veroyatn. i yeye primenen.*, **14**, No. 4, 639–653 (1969).
50. Tarakanov, V. Ye., and V. P. Chistyakov, On the cyclic structure of random permutations, *Matem. sbornik,* **96** (138), No. 4, 594–600 (1975).
51. Tumanyan, S. Kh., On the asymptotic distribution of the χ^2 criterion, *Dokl. Akademii Nauk SSSR,* **94**, No. 6, 1011–1012 (1954).
52. Tumanyan, S. Kh., Asymptotic χ^2 distribution with simultaneous increase in the number of observations and the number of groups, *Teoriya veroyatn. i yeye primenen.*, **1**, No. 1, 131–145 (1956).
53. Wilks, S. S., *Mathematical Statistics,* New York, Wiley, 1962.
54. Feller, W., *Introduction to Probability Theory and its Applications,* Vol. 2, Interscience-John Wiley, New York, 1971.
55. Chistyakov, V. P., Computation of the power of the empty cell test, *Teoriya veroyatn. i yeye primenen.*, **9**, No. 4, 718–724 (1964).
56. Chistyakov, V. P., Discrete limit distributions in the problem of shot with arbitrary probabilities of occupancy of boxes, *Matem. Zametki,* **1**, No. 1, 9–16 (1967).
57. Balakrishnan, V., G. Sankaranarayanan, and C. Suyambulingom, Ordered cycle lengths in a random permutation, *Pacif. J. Math.*, **36**, No. 3, 603–613 (1971).
58. Barton, D. E., and F. N. David, *Combinatorial Chance,* London, Griffin, 1962.
59. Baum, L. E. and P. Billingsley, Asymptotic distributions for the coupon collector's problem, *Ann. Math. Stat.*, **36**, No. 6, 1835–1839 (1965).
60. Békéssy, A., On classical occupancy problems. I, *Magy. tud. akad. Mat. kutató int. közl.*, **8**, Nos. 1–2, 59–71 (1963).
61. Békéssy, A., On classical occupancy problems. II. Sequential occupancy, *Magy. tud. akad. Mat. kutató int. Közl.*, **9**, Nos. 1–2, 133–141 (1964).
62. Békéssy, A., A lottójátékkal kapcsolatos néhány cellabetöltési problémáról. I., *Mat. lapok,* **15**, No. 4, 317–329 (1964).
63. Békéssy, A., A lottójátékkal kapcsolatos néhány cellabetöltési problémárol. II., *Mat. lapok,* **16**, No. 1–2, 57–66 (1965).
64. Best, M. R., The distribution of some variables on a symmetric groups, *Proc. Kon. ned. akad. wet.*, **A73**, No. 5, 385–402 (1970).

65. Brayton, R. K., On the asymptotic behavior of the number of trials necessary to complete a set with a random selection, *J. Math. Anal. and Appl.*, **7**, No. 1, 31–61 (1963).
66. Chowla, S., I. N. Herstein, and K. Moore, On recursions with symmetric groups, *Can. J. Math.*, **3**, 328–334 (1951).
67. Csorgo, M., and I. Guttman, On the consistency of the two-sample empty cell test, *Can. Math. Bull.*, **7**, No. 1, 57–63 (1964).
68. Csorgo, M., and I. Guttman, On the empty cell test, *Technometrics*, **4**, No. 2, 235–247 (1962).
69. Darling, D. A., Some limit theorems associated with multinomial trials, *Proc. 5th Berkeley Sympos. Math. Statist. and Probabil.*, 1965–1966, vol. 2, Part 1, Berkeley–Los Angeles, 1967, 345–350.
70. David, F. N., Two combinatorial tests whether a sample has come from a given population, *Biometrika*, **37**, 97–110 (1950).
71. Dixon, D., The probability of generating the symmetric group, *Math. Z.*, **110**, No. 3, 199–205 (1969).
72. Dwass, M., More birthday surprises, *J. Combin. Theory*, **7**, No. 2, 258–261 (1969).
73. Erdös, P., and A. Rényi, On a classical problem of probability theory, *Magy. tud. akad. Mat. kutató int. közl.*, **6**, No. 1–2, 215–220 (1961).
74. Erdös, P., and P. Turan, On some problems of a statistical group-theory. I., *Z. Wahrscheinlichkeitstheor. und verw. Geb.*, **4**, No. 2, 175–186 (1965).
75. Erdös, P., and P. Turan, On some problems of a statistical group-theory. III, *Acta math. Acad. sci. hung.*, **18**, No. 3–4, 309–320 (1967).
76. Hafner, R., Verteilungskonvergenz gegen Poisson-Prozesse von Punktkomplexen im Z^m, *Teoriya veroyatn. i yeye primenen.*, **18**, No. 1, 133–150 (1973).
77. Hafner, R., Asymptotische Normalität der Anzahl zufälliger Moleküle, *Sitzungsber. Osterr. Akad. Wiss. Math.-naturwiss. Kl.*, **2**, 181, No. 4–7, 215–251 (1973).
78. Hafner, R., Neuer Beweis eines klassischen Besetzungsproblems, Sitzungsber. *Osterr. Akad. Wiss. Math.-naturwiss. Kl.*, **2**, 181, No. 8–10, 269–289 (1973).
79. Harris, B., C. J. Park, The distribution of linear combinations of the sample occupancy numbers, *Proc. Kon. ned. akad. wet.*, **A74**, No. 2, 121–134 (1971).

80. Harris, B., C. J. Park, The limiting distribution of the sample occupancy numbers from the multinomial distribution with equal cell probabilities, *Ann. Inst. Statist. Math.*, **23**, No. 1, 125–133 (1971).
81. Holst, L., Asymptotic normality in a generalized occupancy problem, *Z. Wahrscheinlichkeitstheor. und verw. Geb.*, **21**, No. 2, 109–120 (1972).
82. Holst, L., Limit theorems for some occupancy and sequential occupancy problems, *Ann. Math. Stat.*, **42**, No. 5, 1671–1680 (1971).
83. Karlin, S., Central limit theorems for certain infinite urn schemes, *J. Math. and Mech.*, **17**, No. 4, 373–401 (1967).
84. Kitabatake, S., A remark on a non-parametric test, *Math. Jap.*, **5**, No. 1, 45–49 (1958).
85. Markoff, A. A., *Wahrscheinlichkeitsrechnung,* Leipzig and Berlin, 1912.
86. Mises, R., Über Aufteilungs und Besetzungs-Wahrscheinlichkeiten, *Revu de la Faculté des Sciences de l'Université d'Istanbul,* N. S., **4**, 145–163 (1939).
87. Moser, L., and M. Wyman, On the solutions of $x^d = 1$ in symmetric groups, *Can. J. Math.*, **7**, No. 2, 159–168 (1955).
88. Newman, D. J., and L. Shepp, The double dixie cup problem, *Amer. Math. Mon.*, **67**, No. 1, 58–61 (1960).
89. Okamoto, M., On a non-parametric test, *Osaka J. Math.*, **4**, 77–85 (1952).
90. Pólya, G., Eine Wahrscheinlichkeitsaufgabe in der Kundenwerbung, *Zeitschrift für Angewandte Math. und Mech.*, **10**, 1–3, 96–97 (1930).
91. Rényi, A., Three new proofs and generalization of a theorem of Irving Weiss, *Magy. tud. akad. Mat. kutatö int. közl.*, **7**, No. 1–2, 203–214 (1962).
92. Rényi, A., *Wahrscheinlichkeitsrechnung,* VEB Deutscher Verlag der Wissenschaften, Berlin, 1962.
93. Rosen, B., On the central limit theorem for sums of dependent random variables, *Z. Wahrscheinlichkeitstheor. und verw. Geb.*, **7**, No. 1, 48–82 (1967).
94. Rosen, B., Asymptotic normality in a coupon collector's problem, *Z. Wahrscheinlichkeitstheor. und verw. Geb.*, **13**, No. 3–4, 256–279 (1969).
95. Rosen, B., On the coupon collector's waiting time, *Ann. Math. Stat.*, **41**, No. 6, 1952–1969 (1970).

96. Shepp, L. A., and S. P. Lloyd, Ordered cycle lengths in a random permutation, *Trans. Amer. Math. Soc.,* **121**, No. 2, 340–357 (1966).

97. Stecks, G. P., Limit theorems for conditional distributions, *Univ. Calif. Publ. Statist.,* **2**, No. 12, 237–287 (1957).

98. Weiss, I., Limiting distributions in some occupancy problems, *Ann. Math. Stat.,* **29**, No. 3, 878–884 (1958).

99. Wilks, S. S., A combinatorial test for the problem of two samples from continuous distributions, *Proc. 4th Berkeley Sympos. Math. Stat. and Probability,* 1960, vol. 1, Berkeley–Los Angeles, 1961, 707–717.

INDEX